Hydrogels - Smart Materials Shaping Biomedical, Robotic, and Neuroelectronic Frontiers

Edited by Xiao-Feng Sun

Published in London, United Kingdom

Hydrogels - Smart Materials Shaping Biomedical, Robotic, and Neuroelectronic Frontiers
http://dx.doi.org/10.5772/intechopen.1010351
Edited by Xiao-Feng Sun

Assistants to the Editor: Muhammad Sheraz

Contributors
Daisuke Ishikawa, Elham Farahani, Mahsa Heydari, Maryam Shamshiri, Mojtaba Taghipoor, Muhammad Sheraz, Nafiseh Dorosti, Nazia Mustary Nayna, Roujun Chen, Sabrina Sabnam Samonty, Sima Nematipour, Xiao-Feng Sun, Yaqin Lin, Zhi Zheng

First published in London, United Kingdom, 2026 by IntechOpen
IntechOpen is the global imprint of INTECHOPEN LIMITED, registered in England and Wales, registration number: 11086078, 167-169 Great Portland Street, London, W1W 5PF, United Kingdom

For EU product safety concerns: IN TECH d.o.o., Prolaz Marije Krucifikse Kozulić 3, 51000 Rijeka, Croatia, info@intechopen.com or visit our website at intechopen.com.

British Library Cataloguing-in-Publication Data
A catalogue record for this book is available from the British Library

Hydrogels - Smart Materials Shaping Biomedical, Robotic, and Neuroelectronic Frontiers
Edited by Xiao-Feng Sun
p. cm.
Print ISBN 978-1-80631-022-7
Online ISBN 978-1-80631-021-0
eBook (PDF) ISBN 978-1-80631-023-4

Meet the editor

Dr. Xiao-Feng Sun obtained his Ph.D. degree at Bangor University in the United Kingdom in 2006, and then, as a research officer, he worked on an industrial-based project. In 2007, he returned to China and focused his studies on environmental materials at Zhejiang University. Now he works at Northwestern Polytechnical University, and he has published more than 100 papers, which have been cited more than 6500 times by other scientists. His research focuses on natural polymers and nanomaterials for environmental protection and biomedical applications. He has received 8 academic awards from the Chinese government and international societies, and he is an IAAM fellow, a 2020 Highly Cited Chinese Researcher, and a World top 2% scientist.

Contents

Preface

In the quiet intersection of materials science, biology, and engineering, hydrogels have emerged as silent revolutionaries. These water-swollen polymer networks, once overlooked as simple hydrated matrices, now stand at the forefront of technological and medical innovation, bridging the gap between rigid machinery and living systems with their tissue-like softness, biocompatibility, and responsive intelligence. This book is born from a desire to unravel the potential of hydrogels, tracing their journey from fundamental chemistry to transformative applications that touch nearly every aspect of human life.

Hydrogels' magic lies in their duality: they are both structure and function, passive scaffold and active responder. A hydrogel can mimic the elasticity of cartilage, deliver life-saving drugs with pinpoint precision, conduct neural signals without irritating delicate brain tissue, or clean pollutants from water with eco-friendly efficiency. As we delve into the pages ahead, we explore this versatility through four core pillars: foundational science, advanced fabrication, specialized applications, and future frontiers. From the sustainable synthesis of lignin-based networks to programmable architectures of DNA-based hydrogel, from the stimuli-responsive mechanisms that drive hydrogel actuators to hydrogel engineering interventions for precision cartilage regeneration and the brain's electrical symphony, each chapter unpacks the science behind these materials while grounding it in real-world impact.

What makes hydrogels truly extraordinary is their ability to adapt—to swell or shrink, heal or degrade, conduct or insulate—all in response to their environment. This adaptability has opened doors to innovations once confined to science fiction: soft robots that handle fragile biological samples, implantable devices that dissolve after healing, wearable sensors that monitor health in real time, and regenerative scaffolds that guide damaged tissues back to wholeness. However, despite their promise, hydrogels face profound challenges—mechanical fragility, dehydration, and scalability—that demand interdisciplinary ingenuity to overcome. This book does not shy away from these hurdles; instead, it frames them as invitations for progress, highlighting how researchers are merging nanotechnology, 3D printing, synthetic biology, and artificial intelligence to push hydrogels beyond their limits.

Our book, spanning these pages, is guided by a simple yet powerful question: How can a material so soft, so water-rich, and so biologically similar to living tissue redefine what technology can be? Whether you are a student taking your first steps in materials science, a researcher pushing the boundaries of innovation, or a clinician seeking new therapeutic tools, this book offers a comprehensive map of hydrogels' past, present, and future. We celebrate the curiosity of early pioneers who first recognized hydrogels' potential and honor the contemporary scientists whose work continues to expand their capabilities.

As we stand on the cusp of a new era in soft materials, hydrogels remind us that strength need not come from rigidity, and intelligence need not rely on complex circuitry. They embody a more harmonious approach to technology, one that works with nature rather than against it. We hope that this book inspires you to see hydrogels not just as materials, but as partners in solving some of our most pressing challenges, from healing the human body to protecting our planet.

Xiao-Feng Sun
School of Chemistry and Chemical Engineering,
Northwestern Polytechnical University,
Xi'an, China

Chapter 1

Introductory Chapter: Hydrogels

Xiao-Feng Sun

1. Introduction: Overview of smart hydrogels

Hydrogels are semi-solid, soft, viscoelastic materials composed of three-dimensional (3D) networks that are cross-linked with polymeric chains, exhibiting both solid and liquid-like behaviour. These materials are swollen with either aqueous or non-aqueous solvents, which makes them appropriate for many applications [1–3]. Hydrogels are amphiphilic or hydrophilic networks that are connected and stabilized via many crosslinks, such as chemical (covalent) and physical (non-covalent) interactions [4–6]. Hydrogels have a solid polymer network referred to as a 'mesh structure', which is capable of absorbing and holding large quantities of water and biological fluids without disturbing structural and mechanical integrity, giving them a gel-like consistency. Moreover, hydrogel structures are tunable, biocompatible, biodegradable, and functionally adjustable [7–9]. The polymeric network is surrounded by many hydrophilic groups, such as sulfonic acid (-SO_3H), carboxyl (-COOH), hydroxyl (-OH), amide (-$CONH_2$), and amino (-NH_2) groups, that facilitate water uptake and enhance their physicochemical properties [9–12]. Furthermore, the physicochemical characteristics of hydrogels include adhesiveness, conductivity, toughness, elasticity, polarity, adhesive or anti-adhesive properties depending on the application, and self-healing ability [3, 13].

Hydrogels are broadly classified based on different parameters, such as physical characteristics, origin, source, biodegradation rate, ionic charge, swelling behaviour, cross-linking mechanism, preparation method, and stimuli-responsiveness [11, 15]. **Figure 1** illustrates the classification of hydrogels [14].

2. Hydrogels for biomedical applications

Hydrogels are emerging and versatile materials widely used in the biomedical field over the past few years. Among the variety of drug delivery systems, hydrogel-based drug delivery systems are regarded as one of the most efficient approaches due to their unique physicochemical properties [9]. The composition of drug-containing hydrogels includes a nucleus (drug) and a shell (hydrogel), and these types of drug delivery hydrogels are accessible in the form of capsules, injections, and patches. Hydrogels encapsulate the drug and then release that particular drug toward the targeted site for sustained release. After the hydrogel swells, the drug moves out of the hydrogel by diffusion toward the target [16–18]. Lin et al. [19] explained that smart hydrogels make their way to the lesion sites where the treatment is required, reach there, and respond to physiological conditions such as temperature, pH, and enzymatic changes, release the drug steadily, and improve the efficacy of treatment. This type of

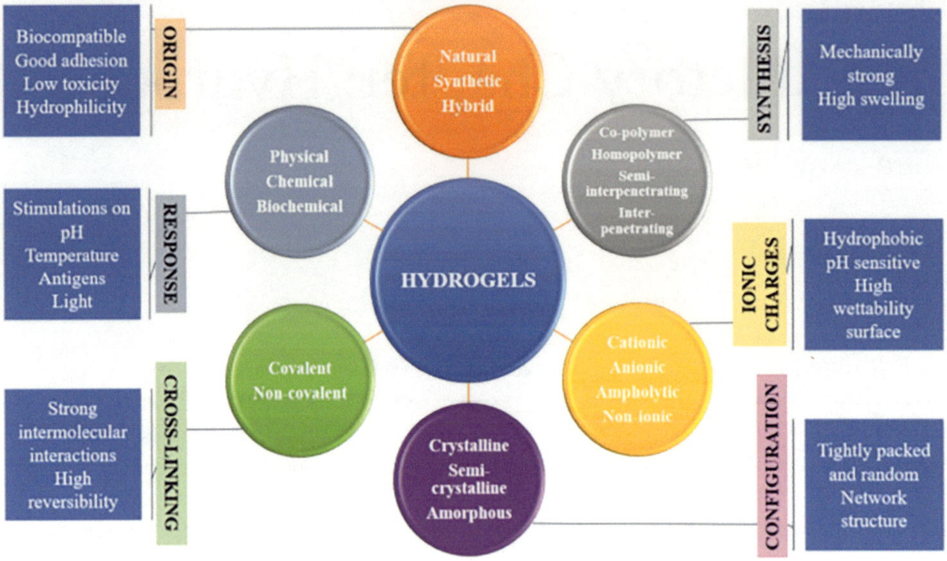

Figure 1.
Classification of hydrogels. Source: Adapted from the study by Pattnaik et al. [14].

drug delivery is suitable for chronic diseases such as tumours and inflammation, where long-term and high-concentration treatment is needed. The movement of the drug molecules from inside to the outside environment is described as external diffusion, which is affected by concentration changes and physiological conditions. The porosity of hydrogels plays a vital role in drug loading and releasing, and they are also used in self-cleaning and wound healing through their moisture contribution to the wound environment. They are broadly used to treat burns, chronic wounds, venous ulcers, and pressure sores. Due to their barrier-forming mechanism and bioadhesive properties, they are potentially used for stitching and surgical stapling. Wound healing has four stages: clotting, cleaning, growth, and maturation stages. Chemically cross-linked hydrogels composed of carboxylated agarose and tannic acid are ideal for wound healing [15, 20]. Due to the high absorption capacity of alginate-based hydrogels, they are employed in speeding up wound healing as they reduce wound exudate and keep the environment wet. The combination of alginate with hyaluronan (composite hydrogel) is effective in enhancing the healing process [21]. Delayed wound healing beyond three months leads to chronic wounds due to prolonged inflammation, and the use of hydrogel dressings to treat them remains effective as they are specially designed with growth factors, biomolecules, and drugs. In addition to this, the addition of phenolic groups from plant or honey sources reduces inflammation and enhances wound recovery [22].

3. Hydrogels for soft robotics

The concept of programmable industrial robotics was first introduced in the twentieth century with the development of the industrial robot Unimate, followed by redundant and continuum robotics [23]. At the start of the twenty-first century, soft robots emerged and gained popularity because of the integration of multiple cutting-edge manufacturing technologies. This growth is attributed to the flexibility of silicone materials and their combination of multiple actuation methods, including tendon-driven [24, 25], electrically-driven, fluid-driven [26], and magnetically-driven techniques [27, 28]. As the applications of silicone-based robotics expanded, their limitations became evident, particularly regarding environmental response and biocompatibility. Researchers then started to study alternative materials in order to overcome this issue. Simultaneously, smart hydrogel materials, known for their excellent features such as responsiveness to external stimuli and biocompatibility, gained tremendous attention in many fields, namely material science, medicine, environmental chemistry, and biochemistry [29, 30]. These unique materials gained appreciation from researchers specializing in soft robotics and smart hydrogels, which ultimately resulted in the development of hydrogel-based continuum soft robotics [31]. As their unique physical characteristics were recognized, researchers began exploring their functions as smart hydrogels, including their attributes of response to external stimuli such as temperature, pH, magnetic field, and electric field [29, 32–34]. The current study of hydrogels aims to make them more intelligent by integrating hydrogels with multiple smart functions, thereby advancing the field of hydrogel-based soft robotics. These advancements include effective drug delivery, shape-changing in narrow spaces, underwater robotics applications, environmental sustainability, and the circular economy [35, 36]. Heiden et al. [37] synthesized a hydrogel-based soft robot using raw material from industrial waste. Along with this, soft robotics also demonstrates its potential by combining sensors and actuators in one system [38, 39] which are effective in various fields such as biomedical applications, including precise surgeries and targeted drug delivery [40], and environmental applications such as recycling and agriculture [41, 42]. Humanoid and robotic dogs made of rigid structures can pose a risk in human–robot interaction, whereas hydrogel-based robots are safer for human–robot interaction [28].

4. Hydrogels for neuroelectronics

Although considerable advancements have been achieved in functional material science, the recording of quality electroencephalogram (EEG) signals, which are essential for the exploration of brain activity, remains a major challenge, particularly through invasive methods. The EEG electrodes are mostly rigid and do not match the properties of human tissue; therefore, they are limited in recording EEG signals. Furthermore, these electrodes are not resistant to environmental noise, making the recording of premium-quality signals difficult in a skin-mounted configuration. Hence, there is a requirement to fabricate interfacial substances that not only increase the performance of EEG electrodes via non-invasive methods but are also compatible with the human body. Due to the soft, viscoelastic, and biocompatible properties of hydrogels, they are considered ideal materials to fulfil this requirement of neuroelectronics, as they can adhere to the skin to improve interfacial contact. Moreover, hydrogels also exhibit a particular frequency effect, known as the damping effect, which helps to decrease surrounding noise and interference during signal recording. Hydrogel-based electrodes are efficient enough to record signals with high quality even in a noise-prone environment. Therefore, these electrodes can

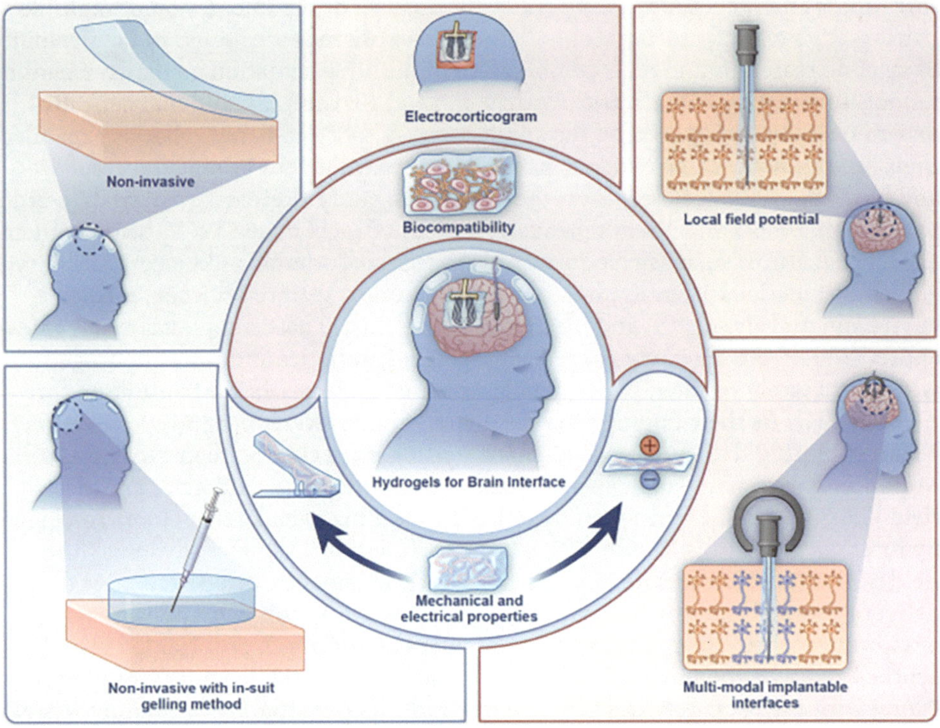

Figure 2.
Application of hydrogels for brain signal monitoring. Source: Adapted from the study by Zhang et al. [43].

be used in combination with computed tomography (CT), magnetic resonance imaging (MRI), and optogenetics for multimodal signal recording. This approach plays a vital role in neurological disease diagnosis as well as in brain science. **Figure 2** demonstrates the application of hydrogels for brain signal monitoring [43].

Author details

Xiao-Feng Sun
School of Chemistry and Chemical Engineering, Northwestern Polytechnical University, Xi'an, China

*Address all correspondence to: xf001sn@nwpu.edu.cn

References

[1] Buema G, Segneanu AE, Herea DD, Grozescu I. Gels for water remediation: Current research and perspectives. Gels. 2024;**10**(9):585

[2] Thang NH, Chien TB, Cuong DX. Polymer-based hydrogels applied in drug delivery: An overview. Gels. 2023;**9**(7):523

[3] Lee MY, Kim HJ, Kim DH, Cha GD, Koo JH. Emerging roles of hydrogels, organogels, and their hybrids in soft bioelectronics and bioplatforms. NPJ Biosensing. 2025;**2**(1):35

[4] Segneanu AE, Bejenaru LE, Bejenaru C, Blendea A, Mogoşanu GD, Biţă A, Boia ER. Advancements in hydrogels: A comprehensive review of natural and synthetic innovations for biomedical applications. Polymers. 2025;**17**(15):2026

[5] Khandaker T, Anik MAAM, Nandi A, Islam T, Islam MM, Hasan MK, Dhar PK, Latif MA, Hossain MS. Recent progress in gel catalysts: Boosting efficiency for sustainable energy applications. Catalysis Science and Technology. 2025;**15**(5):1357–1389

[6] Toews PM, Velraj A, Bates JS. Stimuli-responsive hydrogels, their mass transfer, intermolecular interactions, and applications in biomedical devices. Journal of Materials Science: Materials in Engineering. 2025;**20**(1):66

[7] Rajendran R, Thomas S. In: Handbook of Nutraceuticals: Science, Technology and Engineering Cham, Switzerland: Springer International Publishing, 2026. p. 849–893

[8] Pablos JL, Lozano D, Manzano M, Vallet-Regí M. Regenerative medicine: Hydrogels and mesoporous silica nanoparticles. Materials Today Bio. 2024;**29**:101342

[9] Mehta P, Sharma M, Devi M. Hydrogels: An overview of its classifications, properties, and applications. Journal of the Mechanical Behavior of Biomedical Materials. 2023;**147**:106145

[10] Khan F, Atif M, Haseen M, Kamal S, Khan MS, Shahid S, Nami SAA. Synthesis, classification and properties of hydrogels: Their applications in drug delivery and agriculture. Journal of Materials Chemistry B. 2022;**10**(2): 170–203

[11] Dodda JM, Bezuidenhout D, Yeh YC. Hydrogels: Definition, history, classifications, formation, constitutive characteristics, and applications. In: Dodda JM, Deshmukh K, Bezuidenhout D, editors. Multicomponent Hydrogels Smart Materials for Biomedical Applications. London, UK: The Royal Society of Chemistry; 2023. p. 1–25

[12] Theus AS, Ning L, Kabboul G, Hwang B, Tomov ML. 3D bioprinting of nanoparticle-laden hydrogel scaffolds with enhanced antibacterial and imaging properties. Iscience. 2022;**25** (9):104947

[13] Yuan Y, Zhang Q, Lin S, Li J. Water: The soul of hydrogels. Progress in Materials Science. 2025;**148**:101378

[14] Pattnaik A, Ghosh P, Poonia AK. An overview on advancements in hydrogels for effective wastewater

treatment. Journal of Molecular Liquids. 2025;**424**:127120

[15] Kumar A, Pandey S, Kumar K, Krishnamoorthi S, Tungala K. Hydrogels: Classification, cross-linking methods, characteristics, and current trends in biomedical applications. Polymer Bulletin. 2026;**83**(2):49

[16] Pirsa S, Khodaei SM, Karimi Sani I, Ghasemi Y, Jawhar ZH, Eghbaljoo H. Hydrogels and biohydrogels: Investigation of origin of production, production methods, and application. Polymer Bulletin. 2023;**80**(10): 10593–10632

[17] Lei L, Bai Y, Qin X, Liu J, Huang W, Lv Q. Current understanding of hydrogel for drug release and tissue engineering. Gels. 2022;**8**(5):301

[18] Raina N, Pahwa R, Bhattacharya J, Paul AK, et al. Drug delivery strategies and biomedical significance of hydrogels: Translational considerations. Pharmaceutics 2022;**14**:574

[19] Zainab F, Mir S, Khan SW, Awwad NS, Ibrahium HA. Silane-crosslinked graphene oxide reinforced chitosan/sodium alginate hydrogel for controlled release of insulin. RSC Advances. 2025;**15**(25):19983–20005

[20] Firlar I, Altunbek M, McCarthy C, Ramalingam M, Camci-Unal G. Functional hydrogels for treatment of chronic wounds. Gels. 2022;**8**(2):127

[21] Kapil P, Verma D, Pradhan R, Kalkal A, Packirisamy G. A bioinspired porous and electroactive reduced graphene oxide hydrogel based biosensing platform for efficient detection of tumor necrosis factor-α. Journal of Materials Chemistry B. 2024;**12**(47):12207–12219

[22] Xin C, Cheng Z, Liu W, Li W, Zhu H. The antibacterial and hemostatic activity of Gastrodia elata polysaccharide-based hydrogel embedded with drug-carrying microspheres accelerates diabetic wound healing. Chemical Engineering Journal. 2024;**492**:152403

[23] Wang H, Du J, Mao Y. Hydrogel-based continuum soft robots. Gels. 2025;**11**(4):254

[24] Russo M, Sadati SMH, Dong X, Mohammad A, Walker ID, Bergeles C, Xu K, Axinte DA. Continuum robots: An overview. Advanced Intelligent Systems. 2023;**5**(5):2200367

[25] Wang H, Du J, Mao Y. Cosserat rod-based tendon friction modeling, simulation, and experiments for tendon-driven continuum robots. Micromachines. 2025;**16**(3):346

[26] Pagoli A, Chapelle F, Corrales-Ramon JA, Mezouar Y, Lapusta Y. Review of soft fluidic actuators: Classification and materials modeling analysis. Smart Materials and Structures. 2022;**31**(1):013001

[27] Kim Y, Zhao X. Magnetic soft materials and robots. Chemical Reviews. 2022;**122**(5):5317–5364

[28] Wang H, Mao Y, Du J. Continuum robots and magnetic soft robots: From models to interdisciplinary challenges for medical applications. Micromachines. 2024;**15**(3):313

[29] Yuk H, Wu J, Zhao X. Hydrogel interfaces for merging humans and machines. Nature Reviews Materials. 2022;7(12):935–952

[30] Li W, Liu J, Wei J, Yang Z, Ren C, Li B. Recent progress of conductive

hydrogel fibers for flexible electronics: Fabrications, applications, and perspectives. Advanced Functional Materials. 2023;**33**(17):2213485

[31] Jiang J, Xu S, Ma H, Li C, Huang Z. Photoresponsive hydrogel-based soft robot: A review. Materials Today Bio. 2023;**20**:100657

[32] El-Husseiny HM, Mady EA, Hamabe L, Abugomaa A, et al. Smart/stimuli-responsive hydrogels: Cutting-edge platforms for tissue engineering and other biomedical applications. Materials Today Bio. 2022;**13**:100186

[33] Zhu T, Ni Y, Biesold GM, Cheng Y, Ge M, Li H. Recent advances in conductive hydrogels: Classifications, properties, and applications. Chemical Society Reviews. 2023;**52**(2):473–509

[34] Chen L, Liu F, Abdiryim T, Liu X. Stimuli-responsive hydrogels as promising platforms for soft actuators. Materials Today Physics. 2024;**40**:101281

[35] Ahmaruzzaman M, Roy P, Bonilla-Petriciolet A. Polymeric hydrogels-based materials for wastewater treatment. Chemosphere. 2023;**331**:138743

[36] Huang SC, Zhu YJ, Huang XY, Xia XX, Qian ZG. Programmable adhesion and morphing of protein hydrogels for underwater robots. Nature Communications. 2024;**15**(1):195

[37] Heiden A, Preninger D, Lehner L, Baumgartner M, et al. 3D printing of resilient biogels for omnidirectional and exteroceptive soft actuators. Science Robotics 2022;**7**:eabk2119

[38] Zhang S, Wei H, Tang J. Self-sensing magnetic actuators of bilayer hydrogels. International Journal of Smart and Nano Materials. 2023;**14**(4):496–509

[39] Yang J, Huang W, Peng K, Cheng Z, Lin L, Yuan J, Sun Y, Cho NJ. Versatile agar-zwitterion hybrid hydrogels for temperature self-sensing and electro-responsive actuation. Advanced Functional Materials. 2024;**34**(19):2313725

[40] Rana H, Rana J, Sareen D, Goswami S. Value addition to agro-industrial waste through pectin extraction: Chemometric categorization, density functional theory analysis, rheology investigation, optimization using response surface methodology and prospective applications through hydrogel preparation. Journal of Polymers and the Environment. 2024;**32**(7):2965–2987

[41] Ding X, Wang S, Dai R, Chen H, Shan Z. Hydrogel beads derived from chrome leather scraps for the preparation of lightweight gypsum. Environmental Technology & Innovation. 2022;**25**:102224

[42] Vijayaraghavan R, Unnikrishnan G. Preparation of hydrogel from agriculture waste for the improvement of soil irrigation system. The Journal of Solid Waste Technology and Management. 2022;**48**(2):208–216

[43] Zhang M, Hao M, Liu B, Chen J, Ren G, Zhao Y, et al. Recent progress of hydrogels in brain-machine interface. Soft Science 2024;**4**:39

Chapter 2

Lignin-Based Hydrogels: Synthesis, Applications and Future Perspectives

Muhammad Sheraz and Xiao-Feng Sun

Abstract

Lignin-based hydrogels represent a new generation of bio-composite materials that utilize environmentally sustainable resources and exhibit significantly advanced functional properties. As a highly abundant yet underutilized natural polymer, lignin imparts hydrogels with exceptional mechanical strength, biocompatibility, and responsiveness to environmental stimuli. This review discusses various synthetic approaches that harness lignin's unique properties for hydrogel synthesis, including graft polymerization, crosslinking copolymerization, and interpenetrating polymer networks. Each strategy offers distinct advantages and enables the customization of physicochemical properties to meet specific application requirements. The multifunctional nature of lignin-based hydrogels has proven valuable in a range of fields. In biomedicine, they are explored for applications such as wound healing, drug delivery, and tissue engineering. In environmental remediation, they show strong potential for the adsorption of heavy metals and dyes. Despite their promise, challenges remain in the synthesis of lignin-based hydrogels due to the structural heterogeneity and variable chemical reactivity of lignin. Ongoing developments in chemical modification and hybrid hydrogel systems aim to overcome these limitations. Overall, lignin-based hydrogels offer a promising, eco-friendly alternative that supports the green materials agenda across both healthcare and environmental sectors.

Keywords: lignin-based hydrogels, bio-composite materials, graft polymerization, crosslinking copolymerization, interpenetrating polymer networks, biomedicine, environmental remediation

1. Introduction

Hydrogels are polymer-based substances that absorb water and swell [1, 2]. They have received tremendous attention within the scientific community on the one hand and in the industrial domain on the other, due to their unique ability to retain vast amounts of water while still maintaining their form. Because of their exceedingly porous structure, adjustable mechanical properties, and biocompatibility, hydrogels are suitable for all kinds of applications, especially in biomedicine [3, 4], agriculture

[5], food packaging [6], and environmental remediation [7]. Conventional hydrogels are generally synthesized from synthetic polymers that are very effective but raise the question of sustainability and may lack any biological function [8, 9]. With the whole world moving toward greener, eco-friendly technologies, the need for sustainable, renewable, multifunctional hydrogel materials has hence escalated [10, 11]. In this context, research has been directed toward incorporating lignin into hydrogel matrices to improve both the hydrogel profile and functional capabilities.

Lignin, an aromatic, amorphous biopolymer derived from the secondary cell walls of vascular plants, stands out as a highly promising candidate for incorporation into hydrogel systems [12]. It is the second-largest natural polymer on Earth, after cellulose, and constitutes a significant amount of waste generated by the pulp and paper industries [13, 14]. Despite its widespread availability, lignin has historically been underutilized. In most cases, it is either discarded or incinerated as waste. However, developments in lignin valorization have revealed that lignin has an extraordinary capacity to be exploited as a bio-based building block in material science developments [15]. It is composed of aromatic alcohols – p-coumaryl, coniferyl, and sinapyl (**Figure 1**) – which are categorized by their methoxy group count, enabling strong crosslinking. With an abundance of phenolic and aliphatic hydroxyl groups, lignin can be chemically modified through many pathways, thereby making it amenable to hydrogel network formation via a variety of polymerization and crosslinking techniques [16–18]. It also imparts antimicrobial, antioxidant, and UV-blocking characteristics, which are desired by advanced fields for biomedical and environmental uses [19–21]. Its structural complexity also allows for functional tailoring, giving researchers the flexibility to engineer hydrogels with specific performance [22, 23].

The integration of lignin into hydrogel networks is considered a significant advancement in the development of multifunctional and sustainable soft materials [24, 25]. This approach not only enhances the environmental profile of hydrogel systems by reducing reliance on petrochemical-based monomers but also introduces new physicochemical properties that broaden their range of applications [26, 27].

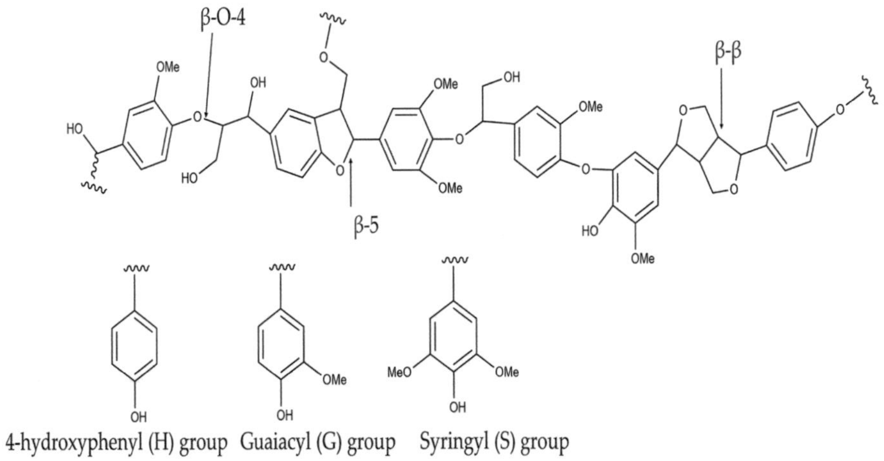

Figure 1.
Lignin structures highlighting the predominant interunit linkages and the associated monomeric units, including 4-hydroxyphenyl (H), guaiacyl (G), and syringyl (S) groups [112].

Lignin-based hydrogels have emerged as a promising research area in recent years due to the unique advantages of lignin. This chapter provides an overview of the synthesis strategies, biomedical and environmental applications, as well as the challenges and future perspectives of lignin-based hydrogels. We examine the various synthetic approaches used to incorporate lignin into hydrogel structures, including graft polymerization and the formation of interpenetrating networks through both covalent and noncovalent crosslinking mechanisms. Furthermore, we highlight the versatile applications of lignin-based hydrogels in the biomedical field – such as wound healing, drug delivery, and tissue engineering – as well as their effectiveness in environmental applications, including the adsorption of heavy metals and organic dyes. Finally, we discuss the current challenges and limitations facing this emerging class of biomaterials, along with future directions aimed at overcoming these barriers to fully exploit the potential of lignin-integrated hydrogel systems.

2. Synthesis approaches

The synthesis of lignin-based hydrogels involves a strategic combination of polymer science and green chemistry to generate multifunctional, sustainable materials. Lignin's rich functional group profile, especially carboxylic, phenolic, and aliphatic hydroxyl groups, makes it highly suitable for chemical modification and integration into polymer networks. The synthesis approaches of lignin hydrogels vary, resulting in differences in application, mechanical strength, swelling behavior, and biocompatibility [12, 28]. Three main strategies are considered in the field: graft polymerization [29], radical polymerization using lignin as a macroinitiator or crosslinker [30, 31], and the synthesis of interpenetrating polymer networks (IPN) [32]. These synthetic routes display their own merits and drawbacks and, therefore, give rise to many types of hydrogel formulations based on particular functional demands.

2.1 Graft polymerization

Graft polymerization is one of the most effective methods in the synthesis of lignin-containing hydrogels [33]. In this method, using free radical initiation, vinyl or acrylic monomers, such as acrylamide, acrylic acid, or N-isopropylacrylamide, are grafted onto a lignin backbone [27, 34]. The hydroxyl groups on lignin serve as anchoring points for the polymer chains, resulting in a network where lignin acts as the backbone or a side-chain scaffold. It, therefore, improves the compatibility of lignin with synthetic polymers and provides precise control over the architecture of the hydrogel. Through thermal initiation, photosensitive initiation, or redox systems, graft copolymerization may be induced, each providing various levels of control over the polymerization kinetics and the final properties of the material [35, 36]. The reason why this method is effective lies in its ability to produce hydrogels that are structurally strong and functionally tunable, with enhanced water retention and mechanical properties [28].

The research by Rajan et al. [37] developed sustainable hydrogels based on lignin, a major waste product from the pulp and paper industry. This was accomplished using a model compound derived from organosolv lignin (OSL) from hybrid poplar.

In the first step, OSL was esterified, effectively grafting 2-hydroxyethyl methacrylate (HEMA) onto the lignin backbone to form OSLH. The OSLH was copolymerized via free radical copolymerization with excess HEMA to form copolymer hydrogels with varying amounts of lignin. **Figure 2** shows the first step, where the OH groups of lignin react with HEMA-Im to form an ester, namely a lignin-HEMA carbonate ester (OSLH). These new hydrogels showed improved water retention, thermal stability, and better mechanical properties than commonly used petrochemical-based materials, presenting a bio-based alternative. The process involves grafting HEMA onto lignin, then copolymerizing the grafted lignin with more HEMA.

2.2 Crosslinking copolymerization

Crosslinking copolymerization is an important approach to forming lignin-based hydrogels with a three-dimensional (3D) network via chemical and/or physical bonding [25, 34, 38]. Covalent crosslinking involves permanent strong bonds (ester, amide, or epoxide linkages) between monomers and functional groups in lignin, which produce hydrogels with composite-like strength and robustness [24, 39, 40]. Alternatively, noncovalent interactions, including hydrogen bonding and π–π stacking, permit flexible, reversible networks that could be suitable for applications requiring self-healing or stimuli-responsiveness [41, 42]. The crosslinking method influences the strength, swelling, and stability of the hydrogel. By tailoring the parameters of polymerization as well as crosslinking, lignin-based hydrogels can be designed for specific functions, including pharmaceutical or biomedical applications, or environmental applications.

The study by Larrañeta et al. [43] presented a straightforward approach to the preparation of lignin-based hydrogels using lignin (LIG), poly(ethylene glycol) (PEG), and poly(methyl vinyl ether-co-maleic acid) (GAN) through an esterification reaction, wherein microwave (MW) radiation can be utilized to accelerate the reaction significantly. **Figure 3a** shows the chemical structure of LIG, GAN, and PEG and displays the proposed cross-linking reaction of the LIG/GAN/PEG hydrogels, in which GAN cross-links with PEG through an esterification reaction. **Figure 3b** shows the FTIR spectra of the carbonyl region of the hydrogels before and after the cross-linking reaction. It is apparent that the carbonyl peak shifts to higher wavenumbers after the cross-linking reaction. The preparation occurs in the solid state, and the subsequent characterization of the hydrogels by

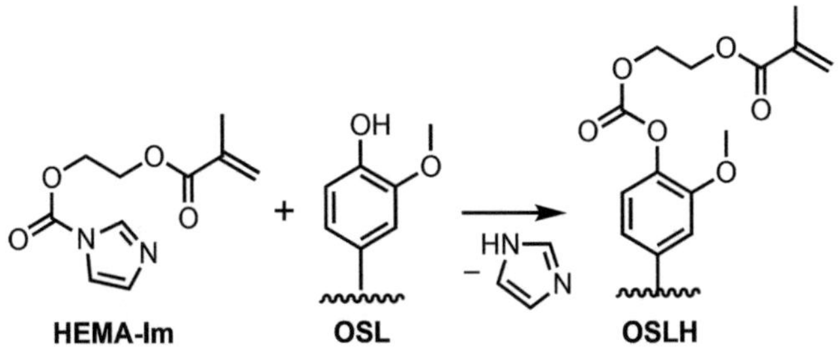

Figure 2.
Modification of lignin (OSL) using HEMA-im to produce OSLH [37].

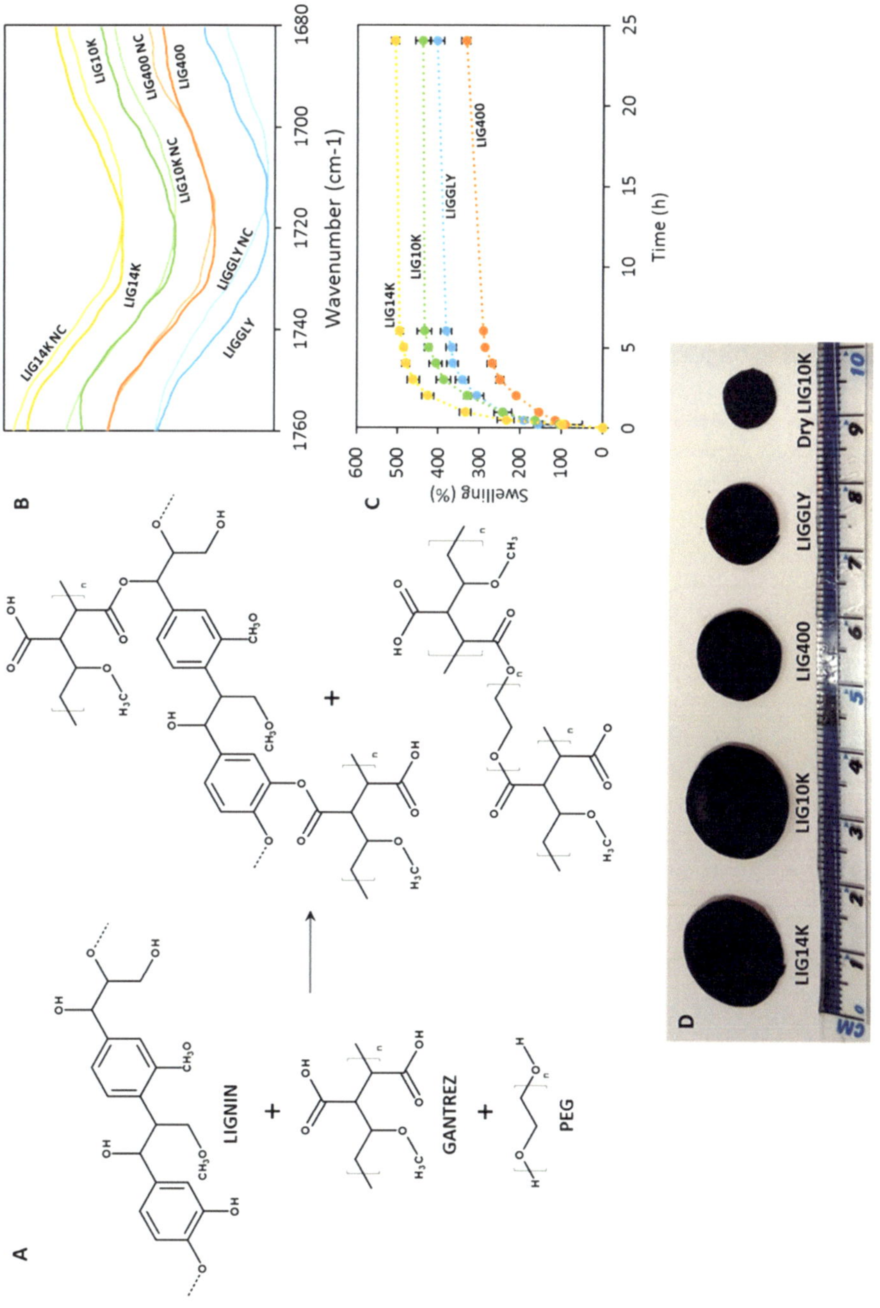

Figure 3.
(a) Chemical structures of LIG, GAN, and PEG, along with proposed crosslinking mechanisms. (b) FTIR spectra focusing on the carbonyl region of LIG/GAN/PEG hydrogels before (NC) and after crosslinking. (c) Swelling behavior of lignin-based hydrogels in PBS over time. (d) Representative images of all swollen lignin-based hydrogels and the dry, unhydrated LIG 10k (n = 3) [43].

solid-state NMR shows that the LIG contents range from 24% to 41%, and the hydrogels have a water uptake of up to 500%, as shown in **Figure 3c**. **Figure 3d** shows representative images of all swollen lignin-based hydrogels and the dry, unhydrated LIG.

2.3 Interpenetrating polymer networks

This method involves the development of IPNs, in which lignin or polymers based on lignin are physically entangled or chemically linked with the other polymer network. IPNs can be made by swelling a preformed polymer in a monomer solution and subsequently polymerizing it in situ, or by polymerizing two monomer systems in parallel with lignin [32, 44, 45]. The resultant hydrogels will have a dual-network structure that combines characteristics from each polymer network, such as rapid healing or biocompatibility from lignin, and increased mechanical strength from each synthetic polymer. This may be a benefit if there are synergistic properties to be served, such as high stretchability, self-healing, or stimuli-responsiveness. Furthermore, it may afford the construction of hydrogels with gradients in composition or spatially separated domains, which may benefit certain complicated energy-related applications [46].

The study by Oveissi et al. [47] examined the fabrication of tough hydrogels derived from inexpensive natural polymers – lignin, which can be incorporated into hydrophilic polyether-type polyurethane (HPU) to produce lignin-HPU hydrogels. Interestingly, the introduction of only 2.5 wt% lignin was enough to enhance the mechanical properties of the HPU hydrogels, quantitatively increasing both the fracture energy and the measured Young's modulus, with no apparent effect on swelling capability. The mimicked lignin-HPU hydrogels exhibited enhanced lap shear adhesiveness, rapid load recoverability, and processability for applications such as 3D printing and fiber spinning. Furthermore, the study confirmed that hydrogen bonding was the primary mechanism contributing to the enhanced toughness of the hydrogels and demonstrated that the developed hydrogels were biocompatible with human dermal fibroblasts. These findings highlight the potential of lignin-HPU hydrogels in a variety of applications, such as wearable devices and biomedical materials.

The study by Naficy et al. [48] produced thin, tough, and pH-sensitive hydrogel films demonstrating rapid load recovery for actuator and sensor applications. The hydrogels were fabricated using a polyether-based polyurethane and poly(acrylic acid) through a simple method, resulting in hydrogel films ranging in thickness between 20 and 570 μm. At the transition point of the hydrogels (pH 4), the hydrogel films exhibited a volume change of less than a factor of two. Each of the hydrogels sustained tensile extensibilities of up to 350% at each pH, respectively, and Young's modulus and tensile strength were compliant with 580–910 kPa and 715–1,320 kPa, respectively, depending on the pH. Unlike double-network hydrogels, these hydrogel films can be stretched to 100% extension and returned to their original length without permanent damage. The results also demonstrated the rapid load recovery aspect of these films, a feature not previously demonstrated with hydrogel films, and a performance characteristic not shared with hybrid ionic-covalent interpenetrating network hydrogels. **Table 1** provides a comparison of synthesis approaches with their advantages and disadvantages.

Approach	Advantages	Disadvantages
Graft polymerization	Strong and tunable, good compatibility, versatile initiation methods	Multistep relies on initiators, heterogeneity issues
Crosslinking copolymerization	Strong (covalent), reversible (noncovalent), tunable swelling, possible green methods	Brittle (overcrosslinked), weak (noncovalent), harsh conditions may be needed
IPNS	Synergistic dual properties: toughness and self-healing, suitable for smart devices	Complex synthesis, with the risk of phase separation, requires careful optimization

Table 1.
Comparison of synthesis approaches with advantages and disadvantages.

3. Applications

3.1 Biomedical application

The hydrogels derived from lignin exhibit multifunctional performance across a myriad of biomedical applications, attributable to their biocompatibility and antioxidant functionality [43, 72]; in addition, the hydrogels provide flexible, moist environments that foster tissue healing [49, 51, 70, 73]. As wound dressings, the hydrogels can maintain a moist environment that protects against infection and promotes tissue regeneration. The benefits of the hydrogels in wound dressings also take advantage of the inherent antimicrobial and antiinflammatory properties of lignin [49, 52, 54, 55, 62, 63]. Specifically, for drug delivery systems, lignin-based hydrogels can provide controlled release of therapeutics based on stimuli, for example, pH or temperature, and can also provide localized and sustained release [58, 61, 67, 74, 75]. In tissue engineering, lignin can be used to enhance structural fidelity and cellular interaction while enabling the development of biodegradable matrices that provide an environment mimicking the natural extracellular process for cellular growth and tissue healing [51, 76, 77]. Lignin-based hydrogels are therefore applicable across a wide array of biomedical applications, making them valuable as multifunctional materials in healthcare. **Table 2** describes the details of lignin-based hydrogels for biomedical applications.

The research by Lin et al. [78] developed a new conductive hydrogel, PLAP, made of PVA, polypyrrole (PPy), lignin sulfonate (Lgs), and silver nanoparticles (Ag NPs) for chronic wound healing. Lgs enhanced the dispersion of PPy in water and enabled conductivity, while Ag NPs provided antibacterial features. The hydrogel showed high conductivity (2.82 ± 0.04 mS/cm), antibacterial ability against *E. coli* and *S. aureus*, biocompatibility, adhesion, and ability to promote hemostasis. In vivo studies showed improved wound closure ($93.71 \pm 2.46\%$ within 14 days) with electrical stimulation (ES), which potentially increased collagen deposition and renewal in epithelial tissue, suggesting PLAP hydrogels are an effective platform for ES-assisted chronic wound healing. **Figure 4** represents the synthesis and use of the PLAP hydrogel as a multifunctional dressing for chronic skin wound healing.

3.2 Environmental remediation

Lignin-based hydrogels can be quite promising as sustainable adsorbents. Lignin hydrogels are very attractive, as they are effective in removing wastewater contaminants such as heavy metals [79–81] and synthetic dyes [82–84] in an

Materials	Method/mechanism of hydrogel preparation	Application	Reference
Lignin/bacterial cellulose/dehydrogenative polymer of coniferyl alcohol	Enzymatic polymerization with solvent-assisted impregnation	Chronic wound healing	[49]
Lignin model polymer dehydrogenated polymer in alginate hydrogel	Blending, ionic crosslinking	Wound treatment	[50]
Alkali lignin/chitosan	Blending	Wound healing	[51]
Lignin/chitosan/polyvinyl alcohol (PVA)	Physical crosslinking via freeze-thaw casting	Wound healing	[52]
Lignin/gold(III) chloride hydrate, silver nitrate/potassium persulfate/poly(acrylic acid)	Covalent crosslinking	Photodynamic therapy	[53]
Lignin/poly(ionic liquids)	Supramolecular interactions	Wound healing	[54]
Lignin/silver nanoparticles/hydroxypropyl cellulose/phenylboric acid	Covalent crosslinking	Wound healing	[55]
Lignin/agarose/silk fibroin solution/zinc chromite ($ZnCr_2O_4$) nanoparticles	Covalent crosslinking	Wound healing	[56]
Lignin/poly(lactic acid) (PLA)	Covalent grafting	Drug delivery	[57]
Alkaline or organosolv lignin/PVA	peroxidation reaction/ hydroxymethylation	Drug delivery	[58]
Lignin-based nanoparticles/thiolated hyaluronic acid/silk fibroin	Mixing	Wound healing	[59]
Lignin/PVA	One-pot reaction	Wound dressing	[60]
Lignin/gelatin/ N-(3-Dimethylaminopropyl)-N′-ethylcarbodiimide hydrochloride	Covalent crosslinking	Drug delivery	[61]
Alkali lignin/chondroitin sulphate sodium salt/PVA/chitosan	Freeze and thaw method	Wound healing	[62]
Cashew apple bagasse lignin	Covalent crosslinking	Wound dressing	[63]
Lignin/Au nanoparticles-decorated halloysite nanotubes/PVA/chitosan	Physical crosslinking via freeze–thaw cycles	Wound healing	[64]
Alkali lignin/poly(ethylene glycol) diglycidyl ether	Covalent crosslinking	Management of atopic dermatitis	[65]
Lignin/epigallocatechin-3-gallate/sodium alginate	Covalent crosslinking	Drug delivery, wound dressing	[66]
Lignin methacrylate-acrylamide-PVA hydrogel	Covalent and physical crosslinking	Drug delivery	[67]
Lignin-containing cellulose nanofiber-based hydrogel	Covalent crosslinking via free radical polymerization	Drug delivery	[68]
Sulfonated lignin/PVA/chitosan	Physical crosslinking via freeze–thaw cycles	Drug delivery	[69]
Demethoxylated lignin /polyacrylamide	Covalent crosslinking via free radical polymerization	Wound healing	[70]
Sodium lignosulfonate-silver nanoparticles/acrylic acid/2-(methacryloyloxy)ethyl dimethyl-(3-sulfopropyl)ammonium hydroxide	Covalent crosslinking via free radical polymerization	Wound dressing	[71]

Table 2.
Lignin-based hydrogels for biomedical applications.

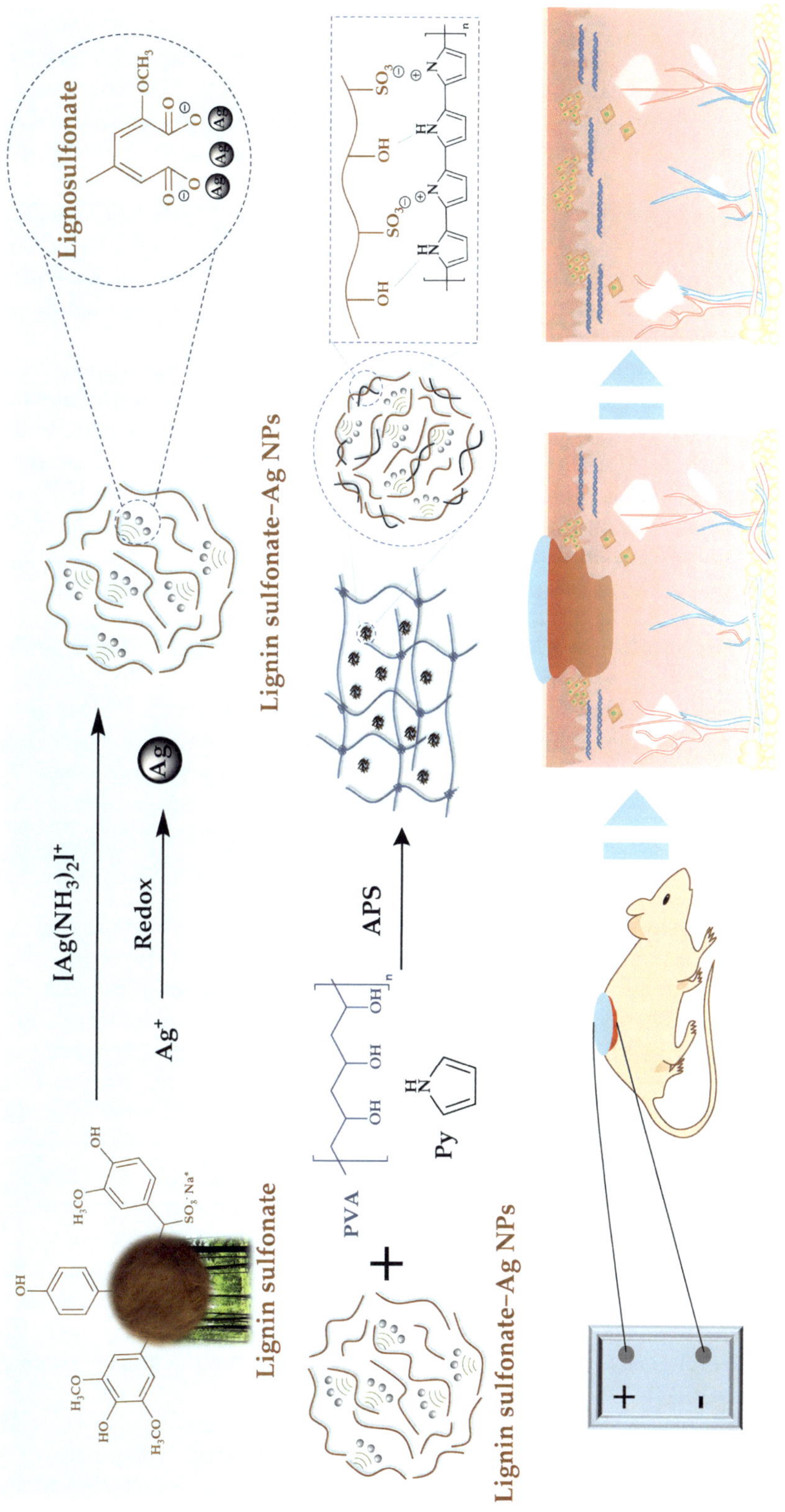

Figure 4.
Schematic diagram depicting the synthesis and multifunctional application of PLAP hydrogel as an advanced dressing for chronic wound healing [78].

environmentally friendly way. Lignin contains a variety of functional groups that chemically bind and adsorb metals through a combination of chelation and ion exchange [85]. The hydrogel enhances these functionalities through a large surface area, high porosity, and extensive swelling, which allow efficient diffusion and capture of contaminants. Along with these characteristics, lignin's aromatic structure allows for π–π interaction and hydrogen bonding with dyes, both cationic [86, 87] and anionic [88, 89]. Thus, lignin-based hydrogels represent a very attractive and biodegradable alternative to the wider range of well-established adsorbents that effectively absorb contaminants in a selective manner, are relatively reusable, and have good mechanical durability. Therefore, they present a sustainable alternative for wastewater treatment. **Table 3** provides details of these lignin hydrogels for the adsorption of metal ions and dyes.

The research by Li et al. [111] investigated the synthesis and adsorption properties of lignin/cellulose hydrogel materials using sodium lignosulfonate and cellulose as raw materials. The hydrogels exhibited a 3D network structure with improved water absorption and retention, showing promise as an adsorbent for dye wastewater treatment. The optimal adsorption conditions for congo red (CR) were determined to be an adsorption time of 4 hours, a pH of 6, and a temperature of 45°C, with a maximum adsorption capacity of 294.0 mg/g, characterized by the Langmuir isothermal model and quasi-secondary kinetic model, indicating systematic monolayer adsorption. The optimal adsorption conditions for malachite green (MG) were an adsorption time of 4 hours, a pH of 4, and a temperature of 60°C, with a maximum adsorption capacity of 129.8 mg/g, consistent with the Freundlich isothermal model and quasi-secondary kinetic model, indicating multilayer adsorption dominated by chemisorption. **Figure 5** shows the preparation of lignin/cellulose hydrogel and the adsorption of methylene blue and malachite green.

4. Challenges and perspectives

Lignin-based hydrogels present several challenges that limit their wider application. One of the most significant issues is the inherent heterogeneity and structural complexity of lignin, which varies depending on its botanical source and extraction method. Such variability strongly influences the reproducibility and consistency of hydrogel properties, including mechanical strength, chemical structure, microstructure, and so on. In addition, lignin's relatively low reactivity and partial miscibility with hydrophilic polymers restrict the efficiency of grafting and copolymerization processes, often requiring premodification steps to improve compatibility. Technical lignin, such as that derived from black liquor in kraft pulping, further demands extensive purification and fractionation to achieve suitable reactivity, thereby increasing the cost and complexity of hydrogel preparation. Even when produced successfully, lignin-based hydrogels can remain mechanically fragile, with limited elasticity and reduced potential for reuse in adsorption–desorption applications such as dye removal or heavy metal ion binding. Therefore, researchers are encouraged to focus on improving the performance of lignin-based hydrogels.

Beyond the intrinsic material challenges, the transition from laboratory-scale synthesis to industrial-scale production introduces additional complexities. Preparation of most lignin-based hydrogels relies on batch polymerization with

Materials	Method/mechanism of hydrogel preparation	Metal ions/dyes	Reference
Lignin/acrylamide/N-isopropyl acrylamide/montmorillonite	Graft copolymerization	Methylene blue	[90]
Sodium lignosulfonate/acrylic acid	Free radical polymerization (ultrasound-assisted)	Cu^{2+}	[91]
Lignin/polyacrylic acid hydrogel	Free radical polymerization	Pb^{2+}	[92]
Lignin/hydroxyethyl celluloses /PVA/ epichlorohydrin	Covalent crosslinking	Rhodamine 6G, crystal violet, and methylene blue	[93]
Lignin/montmorillonite	Redox-initiated free radical polymerization	Cu^{2+}	[94]
Lignin/reduced graphene oxide/ N-methylaniline	Hydrothermal method	Methylene blue, Pb^{2+}	[95]
Alkali lignin/polyacrylic acid	Free radical copolymerization	Pb^{2+}, Cu^{2+}, Cd^{2+}	[96]
Sodium lignosulfonate/acrylamide/ acryloxyethyltrimethylammonium chloride	Free radical graft copolymerization	Acid Red (AR 73)	[97]
Alkali lignin/cellulose nanofibers/carbon dots/polyacrylic acid	Free radical polymerization	Cr^{6+}	[98]
Lignin/2-hydroxyethyl methacrylate / divinylbenzene/PVA	Free radical suspension polymerization	Ni^{2+}	[99]
Sulfomethylated lignin-grafted-polyacrylic acid	Free radical polymerization (ultrasound-assisted)	Ni^{2+}, Co^{2+}, Cd^{2+}, Cu^{2+}, Pb^{2+}	[100]
Lignin nanoparticles-reduced graphene oxide	Physical crosslinking	Methylene blue, Pb^{2+}	[101]
Sulfomethylated lignin/polyacrylic acid	Free radical polymerization	Methylene blue	[102]
Kraft lignin/polyethyleneimine /epichlorohydrin/sodium alginate/$CaCl_2$	Covalent and ionic crosslinking	Methylene blue	[103]
Kraft lignin/PVA/epichlorohydrin	Covalent and physical crosslinking	Methylene blue, crystal violet	[104]
Sodium lignosulfonate/organic montmorillonite/polyacrylic acid	Free radical graft copolymerization	Pb^{2+}, Ni^{2+}, Mn^{2+}	[105]
Lignin/carboxymethyl cellulose/ZnO nanoparticles	Ionic crosslinking	Methylene blue	[106]
Alkali lignin/chitosan/disodium ethylenediaminetetraacetate	Covalent crosslinking	Cu^{2+}, Pb^{2+}	[107]
Alkali lignin/sodium alginate /black liquor	Ionic crosslinking	Malachite green	[108]
Alkali lignin/chitosan hydrogel bead	Covalent crosslinking	Direct Blue-218	[109]
Lignosulfonic acid/poly(sodium 4-styrene sulfonate)	Free radical polymerization	Cu^{2+}	[110]

Table 3.
Lignin-based hydrogels for the adsorption of metal ions and dyes.

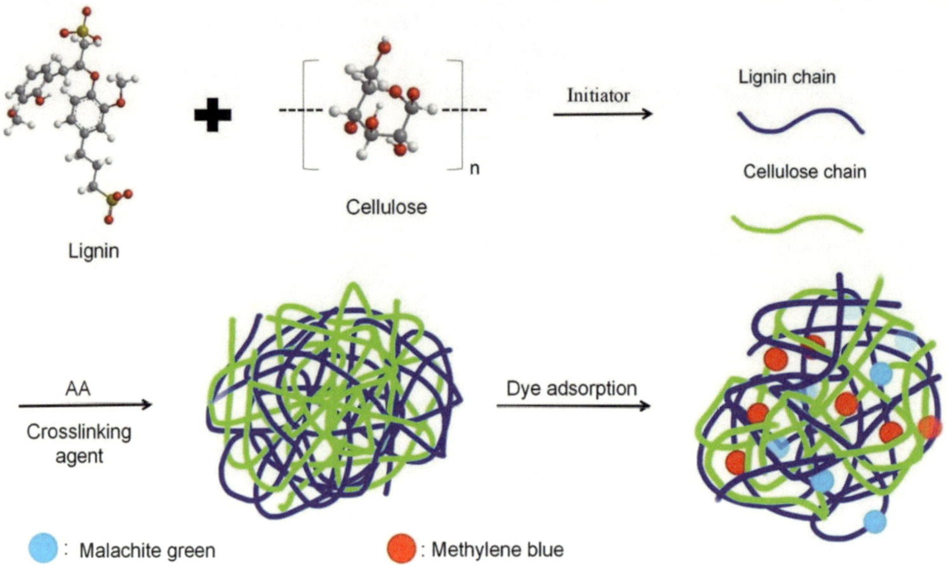

Figure 5.
Preparation of lignin/cellulose hydrogel and adsorption of methylene blue and malachite green [111].

tightly controlled parameters such as temperature, pH, and radical initiation. Replicating these controlled conditions at a large scale is technically difficult, and alternative approaches such as continuous-flow or reactor-based systems require significant optimization to balance cost, efficiency, and safety. Furthermore, lignin feedstocks obtained from industrial processes are highly variable across different pulp and paper mills or bioethanol plants, making it challenging to ensure consistency in large-scale hydrogel production. Industrial adoption, therefore, depends on the establishment of standardized processing methods capable of managing feedstock variability. Another crucial factor is cost competitiveness. Compared to petroleum-derived hydrogels, lignin hydrogels must demonstrate techno-economic feasibility by lowering production costs, especially by capitalizing on the low-cost availability of lignin as a waste product.

Regulatory considerations also represent a critical barrier to the broader application of lignin hydrogels, particularly in biomedical fields. Hydrogels intended for wound healing, drug delivery, or tissue engineering must meet stringent safety standards set by regulatory bodies such as the US Food and Drug Administration (FDA) and the European Medicines Agency (EMA). These requirements include extensive testing of biocompatibility, cytotoxicity, degradation behavior, and long-term stability. Demonstrating reproducibility across batches and ensuring that degradation products are safe will be essential for clinical approval. Even in environmental applications such as wastewater treatment or soil conditioning, lignin hydrogels must undergo regulatory scrutiny to confirm their biodegradability, lack of ecotoxicity, and safety for large-scale use. This regulatory dimension adds another layer of complexity to the industrial translation of lignin hydrogels.

Despite these challenges, lignin hydrogels hold strong promise for future applications as global demand for sustainable and multifunctional materials continues to grow. Advances in chemical modification methods, such as esterification, sulfonation, or etherification, have already improved the reactivity of lignin and its compatibility with synthetic and natural polymers. Hybrid and nanocomposite strategies, such as incorporating lignin with chitosan, cellulose nanofibers, graphene oxide, or metallic nanoparticles, have proven effective in enhancing mechanical strength, conductivity, and antibacterial properties. These modifications open opportunities for advanced applications in biosensing, soft electronics, and biomedical scaffolds. Moreover, the development of stimuli-responsive lignin hydrogels capable of responding to environmental triggers such as pH, temperature, or redox conditions creates exciting prospects for drug delivery systems, wound healing, and smart biomedical devices.

Moving forward, the advancement of lignin-based hydrogels will rely not only on overcoming current technical limitations but also on addressing broader industrial, regulatory, and economic factors. Industrial scalability must focus on the development of efficient, continuous production processes capable of handling variable lignin feedstock without compromising reproducibility. Regulatory compliance requires thorough characterization of biocompatibility and environmental safety, while techno-economic feasibility depends on lowering production costs through the valorization of lignin waste streams. By aligning advances in chemistry, process engineering, and regulatory science, lignin-based hydrogels can evolve from laboratory-scale demonstrations into commercially viable, sustainable materials with wide-ranging environmental and biomedical applications.

5. Conclusions

Lignin's unique chemical structure – characterized by abundant functional groups – confers two key advantages for hydrogel development: first, it supports a broad spectrum of chemical modifications; second, it facilitates integration into diverse hydrogel matrices. Both attributes directly contribute to enhancing critical properties of lignin-based hydrogels, including mechanical strength, antioxidant activity, and environmental responsiveness. Notably, the demonstrated potential of these hydrogels spans both biomedical and environmental applications, which is particularly encouraging. In the biomedical field, representative applications include wound healing, drug delivery, and tissue engineering; in the environmental realm, they encompass wastewater treatment and heavy metal adsorption.

Despite these promising attributes, the practical application of lignin-based hydrogels remains constrained by certain technical barriers – most notably the natural structural variability of lignin and its limited reactivity. However, recent advances in chemical modification strategies, polymer blending techniques, and fabrication methodologies have continuously expanded their practical utility. Currently, research efforts should prioritize investigating two foundational relationships: first, the correlation between lignin's intrinsic structure and the resultant properties of lignin-based hydrogels; second, the structure-performance relationship of the hydrogels themselves. With the growing global emphasis on sustainability and the valorization of industrial by-products, the significance of lignin-based hydrogels is poised to increase further. These materials offer the potential to scale up solutions for pressing biomedical and environmental challenges, aligning with broader goals of circular economy and green technology development.

Acknowledgments

The authors are grateful for the support of Xi'an Science and Technology Plan Project (No. 2024JH-NYYB-0146).

Conflict of interest

The authors declare no conflict of interest.

Author details

Muhammad Sheraz[1,2] and Xiao-Feng Sun[1,2]*

1 School of Chemistry and Chemical Engineering, Northwestern Polytechnical University, Xi'an, P. R. China

2 Shenzhen Research Institute of Northwestern Polytechnical University, Shenzhen, P. R. China

*Address all correspondence to: xf001sn@nwpu.edu.cn

References

[1] Karoyo AH, Wilson LD.A review on the design and hydration properties of natural polymer-based hydrogels. Materials. 2021;**14**(5):1095

[2] Peppas NA, Hoffman AS. Hydrogels. Biomaterials Science: Elsevier. 2020;153–166

[3] Feng W, Wang Z. Tailoring the swelling-shrinkable behavior of hydrogels for biomedical applications. Advanced Science. 2023;**10**(28):2303326

[4] Ding X, Fan L, Wang L, Zhou M, Wang Y, Zhao Y.Designing self-healing hydrogels for biomedical applications. Materials Horizons. 2023;**10**(10): 3929–3947

[5] Kaur P, Agrawal R, Pfeffer FM, Williams R, Bohidar HB.Hydrogels in agriculture: Prospects and challenges. Journal of Polymers and the Environment. 2023;**31**(9):3701–3718

[6] Leyva-Jiménez FJ, Oliver-Simancas R, Castangia I, Rodríguez-García AM, Alañón ME. Comprehensive review of natural based hydrogels as an upcoming trend for food packing. Food Hydrocolloids. 2023;**135**:108124

[7] Zhang J, White JC, Lowry GV, He J, Yu X, Yan C, et al. Advanced enzyme-assembled hydrogels for the remediation of contaminated water. Nature Communications 2025;**16**:3050

[8] Ali A, Ahmed S.Recent advances in edible polymer based hydrogels as a sustainable alternative to conventional polymers. Journal of Agricultural and Food Chemistry. 2018;**66**(27): 6940–6967

[9] Gul K, Gan R-Y, Sun C-X, Jiao G, Wu D-T, Li H-B, et al. Recent advances in the structure, synthesis, and applications of natural polymeric hydrogels. Critical Reviews in Food Science and Nutrition 2022;**62**: 3817–3832

[10] Azeem MK, Islam A, Khan RU, Rasool A, MAuR Q, Rizwan M, et al. Eco-friendly three-dimensional hydrogels for sustainable agricultural applications: Current and future scenarios. Polymers for Advanced Technologies 2023;**34**:3046–3062

[11] Liu Y, Wang J, Chen H, Cheng D. Environmentally friendly hydrogel: A review of classification, preparation and application in agriculture. Science of the Total Environment. 2022;**846**:157303

[12] Pandit S, Sharma P, Prakash A, Lal B, Bhuyan R, Ahmad I, et al. A comprehensive review on technical lignin, lignin hydrogels, properties, preparation, applications & challenges in lab to market transition. Industrial Crops and Products. 2024;**211**:118262

[13] Fadlallah S, Roy PS, Garnier G, Saito K, Allais F.Are lignin-derived monomers and polymers truly sustainable? An in-depth green metrics calculations approach. Green Chemistry. 2021;**23**(4):1495–1535

[14] Yadav V, Kumar A, Bilal M, Nguyen TA, Iqbal HM Lignin removal from pulp and paper industry waste streams and its application. Nanotechnology in Paper and Wood Engineering: Elsevier; 2022. p. 265–283

[15] Boarino A, Klok H-A.Opportunities and challenges for lignin valorization in food packaging, antimicrobial, and agricultural applications. Biomacromolecules. 2023;**24**(3): 1065–1077

[16] Laurichesse S, Avérous L.Chemical modification of lignins: Towards biobased polymers. Progress in Polymer Science. 2014;**39**(7):1266–1290

[17] Kazzaz AE, Feizi ZH, Fatehi P. Grafting strategies for hydroxy groups of lignin for producing materials. Green Chemistry. 2019;**21**(21):5714–5752

[18] Yan L, Huertas-Alonso AJ, Liu H, Dai L, Si C, Sipponen MH. Lignin polymerization: Towards high-performance materials. Chemical Society Reviews. 2025;**54**:6634–6651

[19] Yang D, Fan B, He Y-C. UV-blocking, antibacterial, corrosion resistance, antioxidant, and fruit packaging ability of lignin-rich alkaline black liquor composite film. International Journal of Biological Macromolecules. 2024;**275**:133344

[20] Liu R, Dai L, Xu C, Wang K, Zheng C, Si C.Lignin-based micro-and nanomaterials and their composites in biomedical applications. ChemSusChem. 2020;**13**(17):4266–4283

[21] Cho D-W, Yoon K, Ahn Y, Sun Y, Tsang DC, Hou D, et al. Fabrication and environmental applications of multifunctional mixed metal-biochar composites (MMBC) from red mud and lignin wastes. Journal of Hazardous Materials. 2019;**374**:412–419

[22] Wu Y-C, Wang H-M, Yuan -L-L, Zhang -Q-Q, Liu Y-Q, Shao C-Y, et al. Lignin-based functional hydrogels: An eco-friendly bulk material. ACS Sustainable Chemistry & Engineering 2024;**12**:17952–17976

[23] Kai D, Tan MJ, Chee PL, Chua YK, Yap YL, Loh XJ. Towards lignin-based functional materials in a sustainable world. Green Chemistry. 2016;**18**(5):1175–1200

[24] Gujjala LKS, Kim J, Won W. Technical lignin to hydrogels: An Eclectic review on suitability, synthesis, applications, challenges and future prospects. Journal of Cleaner Production. 2022;**363**:132585

[25] Huo L, Lu Y, Ding W-L, Wang Y, Li X, He H. Recent advances in the preparation, properties, and applications of lignin-based hydrogels and adhesives. Green Chemistry. 2025;**27**:1895–1908

[26] Abbas A, Lai DY, Peng P, She D. Lignin-based functional materials in agricultural application: A review. Journal of Agricultural and Food Chemistry. 2025;**73**(10):5685–5710

[27] Rico-García D, Ruiz-Rubio L, Pérez-Alvarez L, Hernández-Olmos SL, Guerrero-Ramírez GL, Vilas-Vilela JL. Lignin-based hydrogels: Synthesis and applications. Polymers. 2020;**12**(1):81

[28] Mondal AK, Uddin MT, Sujan S, Tang Z, Alemu D, Begum HA, et al. Preparation of lignin-based hydrogels, their properties and applications. International Journal of Biological Macromolecules. 2023;**245**:125580

[29] Kaur R, Sharma R, Chahal GK. Synthesis of lignin-based hydrogels and their applications in agriculture: A review. Chemical Papers. 2021;**75**(9):4465–4478

[30] Diao B, Zhang Z, Zhu J, Li J. Biomass-based thermogelling copolymers consisting of lignin and grafted poly(N-isopropylacrylamide), poly(ethylene glycol), and poly (propylene glycol). RSC Advances. 2014;**4**(81):42996–43003

[31] Shao L, Liu N, Wang Z, Zhan P, Zhang L, Wu Z. Functional lignin-based polymers: Isolation, synthetic methods and high-valued applications. ChemistrySelect. 2023;**8**(33): e202301633

[32] JE PA, Sabino MA.Effect of the presence of lignin or peat in IPN hydrogels on the sorption of heavy metals. Polymer Bulletin. 2010;**65** (5):495–508

[33] Liu H, Chung H.Lignin-based polymers via graft copolymerization. Journal of Polymer Science Part A: Polymer Chemistry. 2017;**55**(21): 3515–3528

[34] Wang Q, Wang B, Chen J, Yang G, Zhang L, Liu K, et al. Lignin-based hydrogel: Mechanism, properties, and applications. Lignin Chemistry: Characterization, Isolation, and Valorization. 2024;433–468

[35] Pourmahdi M, Abdollahi M, Nasiri A. Effect of lignin source and initiation conditions on graft copolymerization of lignin with acrylamide and performance of graft copolymer as additive in water-based drilling fluid. Journal of Petroleum Science and Engineering. 2023;**220**:111253

[36] Zong E, Liu X, Liu L, Wang J, Song P, Ma Z, et al. Graft polymerization of acrylic monomers onto lignin with CaCl2–H2O2 as initiator: Preparation, mechanism, characterization, and application in poly (lactic acid). ACS Sustainable Chemistry & Engineering 2018;**6**:337–348

[37] Rajan K, Mann JK, English E, Harper DP, Carrier DJ, Rials TG, et al. Sustainable hydrogels based on lignin-methacrylate copolymers with enhanced water retention and tunable material properties. Biomacromolecules 2018;**19**:2665–2672

[38] Jin Z, Yu Y, Shao S, Ye J, Lin L, Iiyama K. Lignin as a cross-linker of acrylic acid-grafted carboxymethyl lignocellulose. Journal of Wood Science. 2010;**56**:470–476

[39] Hwang J, Martinez DV, Martinez EJ, Metavarayuth G, Goodlett D, Wang Q, et al. Highly swellable hydrogels prepared from extensively oxidized lignin. Giant. 2022;**10**:100106

[40] Fu C, Ni Y, Chen L, Huang F, Miao Q, Huang L. Design of asymmetric-adhesion lignin-reinforced hydrogels based on disulfide bond crosslinking for strain sensing application. International Journal of Biological Macromolecules. 2022;**212**:275–282

[41] Cao J, Zhao Y, Jin S, Li J, Wu P, Luo Z. Flexible lignin-based hydrogels with self-healing and adhesive ability driven by noncovalent interactions. Chemical Engineering Journal. 2022;**429**:132252

[42] Wang Q, Pan X, Guo J, Huang L, Chen L, Ma X, et al. Lignin and cellulose derivatives-induced hydrogel with asymmetrical adhesion, strength, and electriferous properties for wearable bioelectrodes and self-powered sensors. Chemical Engineering Journal. 2021;**414**:128903

[43] Larrañeta E, Imízcoz M, Toh JX, Irwin NJ, Ripolin A, Perminova A, et al. Synthesis and characterization of lignin hydrogels for potential applications as drug eluting antimicrobial coatings for medical materials. ACS Sustainable Chemistry & Engineering 2018;**6**:9037–9046

[44] Raschip IE, Hitruc EG, Vasile C. Semi-interpenetrating polymer networks containing polysaccharides. II. Xanthan/lignin networks: A spectral and thermal characterization. High Performance Polymers. 2011;**23**(3):219–229

[45] Raschip IE, Vasile C, Ciolacu D, Cazacu G.Semi-interpenetrating polymer networks containing polysaccharides. I Xanthan/Lignin networks. High Performance Polymers. 2007;**19**(5-6):603–620

[46] Zhang H, Zhang W, Ming H, Pang J, Zhang H, Cao G, et al. Design advanced carbon materials from lignin-based interpenetrating polymer networks for high performance sodium-ion batteries. Chemical Engineering Journal. 2018;**341**:280–288

[47] Oveissi F, Naficy S, Le TYL, Fletcher DF, Dehghani F.Tough and processable hydrogels based on lignin and hydrophilic polyurethane. ACS Applied Bio Materials. 2018;**1**(6): 2073–2081

[48] Naficy S, Spinks GM, Wallace GG. Thin, tough, pH-sensitive hydrogel films with rapid load recovery. ACS Applied Materials & Interfaces. 2014;**6** (6):4109–4114

[49] Zmejkoski D, Spasojević D, Orlovska I, Kozyrovska N, Soković M, Glamočlija J, et al. Bacterial cellulose-lignin composite hydrogel as a promising agent in chronic wound healing. International Journal of Biological Macromolecules. 2018;**118**:494–503

[50] Spasojević D, Zmejkoski D, Glamočlija J, Nikolić M, Soković M, Milošević V, et al. Lignin model compound in alginate hydrogel: A strong antimicrobial agent with high potential in wound treatment. International Journal of Antimicrobial Agents 2016;**48**:732–735

[51] Ravishankar K, Venkatesan M, Desingh RP, Mahalingam A, Sadhasivam B, Subramaniyam R, et al. Biocompatible hydrogels of chitosan-alkali lignin for potential wound healing applications. Materials Science and Engineering: C. 2019;**102**:447–457

[52] Zhang Y, Jiang M, Zhang Y, Cao Q, Wang X, Han Y, et al. Novel lignin–chitosan–PVA composite hydrogel for wound dressing. Materials Science and Engineering: C. 2019;**104**:110002

[53] Chandna S, Thakur NS, Kaur R, Bhaumik J.Lignin–bimetallic nanoconjugate doped pH-responsive hydrogels for laser-assisted antimicrobial photodynamic therapy. Biomacromolecules. 2020;**21**(8): 3216–3230

[54] Zhang Y, Yuan B, Zhang Y, Cao Q, Yang C, Li Y, et al. Biomimetic lignin/poly (ionic liquids) composite hydrogel dressing with excellent mechanical strength, self-healing properties, and reusability. Chemical Engineering Journal. 2020;**400**:125984

[55] Deng P, Chen F, Zhang H, Chen Y, Zhou J.Conductive, self-healing, adhesive, and antibacterial hydrogels

based on lignin/cellulose for rapid MRSA-infected wound repairing. ACS Applied Materials & Interfaces. 2021;**13** (44):52333–52345

[56] Eivazzadeh-Keihan R, Aliabadi HAM, Radinekiyan F, Sobhani M, Maleki A, Madanchi H, et al. Investigation of the biological activity, mechanical properties and wound healing application of a novel scaffold based on lignin–agarose hydrogel and silk fibroin embedded zinc chromite nanoparticles. RSC Advances 2021;**11**:17914–17923

[57] Kai D, Ren W, Tian L, Chee PL, Liu Y, Ramakrishna S, et al. Engineering poly (lactide)–lignin nanofibers with antioxidant activity for biomedical application. ACS Sustainable Chemistry & Engineering 2016;**4**:5268–5276

[58] Morales A, Labidi J, Gullón P. Influence of lignin modifications on physically crosslinked lignin hydrogels for drug delivery applications. Sustainable Materials and Technologies. 2022;**33**:e00474

[59] Morena AG, Pérez-Rafael S, Tzanov T.Lignin-based nanoparticles as both structural and active elements in self-assembling and self-healing multifunctional hydrogels for chronic wound management. Pharmaceutics. 2022;**14**(12):2658

[60] Chao Y, Yu S, Zhang H, Gong D, Li J, Wang F, et al. Architecting lignin/ poly (vinyl alcohol) hydrogel with carbon nanotubes for photothermal antibacterial therapy. ACS Applied Bio Materials 2023;**6**:1525–1535

[61] Chiani E, Beaucamp A, Hamzeh Y, Azadfallah M, Thanusha A, Collins MN. Synthesis and characterization of gelatin/lignin hydrogels as quick release drug carriers for Ribavirin. International Journal of Biological Macromolecules. 2023;**224**:1196–1205

[62] Preet J, Pathania K, Kaur J, Singh R, Salunke DB, Pawar SV.A lignin-based biocomposite hydrogel for antimicrobial and wound healing applications. Materials Advances. 2024;**5**(23): 9445–9457

[63] de Albuquerque TL, Cavalcante VGC, da Silva Rocha W, de Macedo AC, Rocha MVP. Hydrogels based on lignin extracted from cashew apple bagasse and its application in antimicrobial wound dressings. International Journal of Biological Macromolecules. 2024;**262**:130169

[64] Xu X, Chao Y, Ma X, Zhang H, Chen J, Zhu J. A photothermally antibacterial Au@ Halloysite nanotubes/lignin composite hydrogel for promoting wound healing. International Journal of Biological Macromolecules. 2024;**258**:128704

[65] Trinh TA, Nguyen TL, Kim J. Lignin-based antioxidant hydrogel patch for the management of atopic dermatitis by mitigating oxidative stress in the skin. ACS Applied Materials & Interfaces. 2024;**16**(26):33135–33148

[66] Shan Z, Jiang B, Wang P, Wu W, Jin Y. Sustainable lignin-based composite hydrogels for controlled drug release and self-healing in antimicrobial wound dressing. International Journal of Biological Macromolecules. 2025;**285**:138327

[67] Singh V, Dhukia S, Khullar L, Sihag S, Kaur R, Harjai K, et al. Synthesis and Characterization of Biomass Derived Modified Lignin Based

PVA Hydrogel as Drug Delivery System for Ciprofloxacin. ChemistrySelect. 2025;**10**(6):e202404597

[68] Cheng G, Zeng F, Liu X, Yang Q, Wei S, Huang Q. Mussel-inspired adhesive and tough hydrogel for drug release based on lignin-containing cellulose nanofiber. International Journal of Biological Macromolecules. 2025;**306**:141458

[69] Wang B, Qiu D, Gu Y, Shan Z, Shi R, Luo J, et al. A lignin-based controlled/sustained release hydrogel by integrating mechanical strengthening and bioactivities of lignin. Journal of Bioresources and Bioproducts 2025;**10**:62–76

[70] Xie C, Jiang S, Li Y, Gan D, Wang Y, Zhang H, et al. Self-oxygenating, anti-freezing, antioxidant, and UV-resistant lignin hydrogels for wound healing at high-altitude. Chemical Engineering Journal. 2025;164422

[71] Jin Z, Gong H, Chen B, Jiang Y, Su Y, Zhou J, et al. Novel functional hydrogels based on lignin-silver nanoparticles with adhesion, antimicrobial, antioxidant and anti-freezing properties for wound dressings and pressure strain sensors. International Journal of Biological Macromolecules. 2025;**291**:138853

[72] Ullah I, Chen Z, Xie Y, Khan SS, Singh S, Yu C, et al. Recent advances in biological activities of lignin and emerging biomedical applications: A short review. International Journal of Biological Macromolecules. 2022;**208**:819–832

[73] Li X, Chen R, Tang X, Chen M, Fan Y. A collagen/nanocellulose/lignin hydrogel dressing mimicking to the plant cell wall for enhanced wound healing and bacterial inhibition. Industrial Crops and Products. 2025;**228**:120872

[74] Marcelo G, López-González M, Trabado I, Rodrigo MM, Valiente M, Mendicuti F. Lignin inspired PEG hydrogels for drug delivery. Materials Today Communications. 2016;7:73–80

[75] Anghel N, Melinte V, Spiridon I, Pertea M.Antioxidant, antimicrobial, and kinetic studies of Β-cyclodextrin crosslinked with lignin for drug delivery. Pharmaceutics. 2022;**14** (11):2260

[76] Kumar R, Butreddy A, Kommineni N, Reddy PG, Bunekar N, Sarkar C, et al. Lignin: Drug/gene delivery and tissue engineering applications. International Journal of Nanomedicine. 2021;**16**:2419–2441

[77] Afewerki S, Wang X, Ruiz-Esparza GU, Tai C-W, Kong X, Zhou S, et al. Combined catalysis for engineering bioinspired, lignin-based, long-lasting, adhesive, self-mending, antimicrobial hydrogels. Acs Nano 2020;**14**: 17004–17017

[78] Lin J, Chen M, Zhao W, Zhang S, Liu J, Zhou Y, et al. Lignin-mediated dual conductive hydrogels with high conductivity, antibacterial activity and biocompatibility for chronic wound repair. Gels 2025;**11**:283

[79] Yao Q, Xie J, Liu J, Kang H, Liu Y. Adsorption of lead ions using a modified lignin hydrogel. Journal of Polymer Research. 2014;**21**:1–16

[80] Ding W, Sun H, Li X, Li Y, Jia H, Luo Y, et al. Environmental applications

of lignin-based hydrogels for Cu remediation in water and soil: Adsorption mechanisms and passivation effects. Environmental Research. 2024;**250**:118442

[81] Tian R, Liu Q, Zhang W, Zhang Y. Preparation of lignin-based hydrogel and its adsorption on Cu 2+ ions and Co 2+ ions in wastewaters. Journal of Inorganic and Organometallic Polymers and Materials. 2018;**28**:2545–2553

[82] Domínguez-Robles J, Peresin MS, Tamminen T, Rodríguez A, Larrañeta E, Jääskeläinen A-S. Lignin-based hydrogels with "super-swelling" capacities for dye removal. International Journal of Biological Macromolecules. 2018;**115**:1249–1259

[83] Jiang M, Niu N, Chen L. A template synthesized strategy on bentonite-doped lignin hydrogel spheres for organic dyes removal. Separation and Purification Technology. 2022;**285**:120376

[84] Wang S, Chen X, Yin Y, Meng H, Wang Y, Xiao Z, et al. Lignin-based hydrogels for efficient dye removal via synergistic effect of multiple interactions. Industrial Crops and Products. 2022;**189**:115840

[85] Zheng L, Seidi F, Wu W, Pan Y, Xiao H. Dual-functional lignin-based hydrogels for sustained release of agrochemicals and heavy metal ion complexation. International Journal of Biological Macromolecules. 2023;**235**:123701

[86] Yu C, Wang F, Zhang C, Fu S, Lucia LA. The synthesis and absorption dynamics of a lignin-based hydrogel for remediation of cationic dye-contaminated effluent. Reactive and Functional Polymers. 2016;**106**:137–142

[87] Mondal AK, Xu D, Wu S, Zou Q, Lin W, Huang F, et al. High lignin containing hydrogels with excellent conducting, self-healing, antibacterial, dye adsorbing, sensing, moist-induced power generating and supercapacitance properties. International Journal of Biological Macromolecules. 2022;**207**:48–61

[88] Meng Y, Li C, Liu X, Lu J, Cheng Y, Xiao L-P, et al. Preparation of magnetic hydrogel microspheres of lignin derivate for application in water. Science of the Total Environment. 2019;**685**:847–855

[89] Wang H, Chen C, Dai K, Xiang H, Kou J, Guo H, et al. Selective adsorption of anionic dyes by a macropore magnetic lignin-chitosan adsorbent. International Journal of Biological Macromolecules. 2024;**269**:131955

[90] Wang Y, Xiong Y, Wang J, Zhang X. Ultrasonic-assisted fabrication of montmorillonite-lignin hybrid hydrogel: Highly efficient swelling behaviors and super-sorbent for dye removal from wastewater. Colloids and Surfaces A: Physicochemical and Engineering Aspects. 2017;**520**:903–913

[91] Wang X, Wang Y, Hou H, Wang J, Hao C.Ultrasonic method to synthesize glucan-g-poly (acrylic acid)/sodium lignosulfonate hydrogels and studies of their adsorption of Cu2+ from aqueous solution. ACS Sustainable Chemistry & Engineering. 2017;**5**(8):6438–6446

[92] Sun Y, Ma Y, Fang G, Li S, Fu Y. Synthesis of acid hydrolysis lignin-g-poly-(acrylic acid) hydrogel superabsorbent composites and adsorption of lead ions. BioResources. 2016;**11**(3):5731–5742

[93] Huang S, Wu L, Li T, Xu D, Lin X, Wu C. Facile preparation of biomass lignin-based hydroxyethyl cellulose super-absorbent hydrogel for dye pollutant removal. International Journal of Biological Macromolecules. 2019;**137**:939–947

[94] Sun X-F, Hao Y, Cao Y, Zeng Q. Superadsorbent hydrogel based on lignin and montmorillonite for Cu (II) ions removal from aqueous solution. International Journal of Biological Macromolecules. 2019;**127**:511–519

[95] Qian H, Wang J, Yan L.Synthesis of lignin-poly (N-methylaniline)-reduced graphene oxide hydrogel for organic dye and lead ions removal. Journal of Bioresources and Bioproducts. 2020;**5** (3):204–210

[96] Liu M, Liu Y, Shen J, Zhang S, Liu X, Chen X, et al. Simultaneous removal of Pb2+, Cu2+ and Cd2+ ions from wastewater using hierarchical porous polyacrylic acid grafted with lignin. Journal of Hazardous Materials. 2020;**392**:122208

[97] Wei S, Chen W, Tong Z, Jiang N, Zhu M.Synthesis of a functional biomass lignin-based hydrogel with high swelling and adsorption capability towards Acid Red 73. Environmental Science and Pollution Research. 2021;**28** (37):51306–51320

[98] Yuan H, Peng J, Ren T, Luo Q, Luo Y, Zhang N, et al. Novel fluorescent lignin-based hydrogel with cellulose nanofibers and carbon dots for highly efficient adsorption and detection of Cr (VI). Science of the Total Environment. 2021;**760**:143395

[99] Goliszek M, Kołodyńska D, Pylypchuk IV, Sevastyanova O, Podkościelna B. Synthesis of lignin-containing polymer hydrogels with tunable properties and their application in sorption of nickel (II) ions. Industrial Crops and Products. 2021;**164**:113354

[100] Jiao G-J, Ma J, Li Y, Jin D, Zhou J, Sun R. Removed heavy metal ions from wastewater reuse for chemiluminescence: Successive application of lignin-based composite hydrogels. Journal of Hazardous Materials. 2022;**421**:126722

[101] Sohni S, Hassan T, Khan SB, Akhtar K, Bakhsh EM, Hashim R, et al. Lignin nanoparticles-reduced graphene oxide based hydrogel: A novel strategy for environmental applications. International Journal of Biological Macromolecules. 2023;**225**:1426–1436

[102] Sun S-F, Wan H-F, Zhao X, Gao C, Xiao L-P, Sun R-C. Facile construction of lignin-based network composite hydrogel for efficient adsorption of methylene blue from wastewater. International Journal of Biological Macromolecules. 2023;**253**:126688

[103] Wang C, Feng X, Shang S, Liu H, Song Z, Zhang H. Lignin/sodium alginate hydrogel for efficient removal of methylene blue. International Journal of Biological Macromolecules. 2023;**237**:124200

[104] Jung S, Yun H, Kim J, Kim J, Yeo H, Choi I-G, et al. Lignin/PVA hydrogel with enhanced structural stability for cationic dye removal. International Journal of Biological Macromolecules. 2024;**257**:128810

[105] Ren H, Hao Y, Sheng J, Zhang G, Zheng J, Ti X, et al. A lignin-based

composite hydrogel for the removal of heavy metal ions from aqueous solution. Journal of Wood Chemistry and Technology 2024;**44**:277–291

[106] Fu J, Yap JX, Leo CP, Chang CK. Enhanced photocatalytic regeneration of carboxymethyl cellulose/lignin/ZnO complex hydrogel after methylene blue adsorption. International Journal of Biological Macromolecules. 2024;**274**:133510

[107] Zhao X, Zhao H, Mei A, Peng L, Sun J. Novel chitosan/lignin hydrogel prepared by the Mannich reaction for Pb (II) and Cu (II) removal from aqueous solution. International Journal of Biological Macromolecules. 2025;**285**:138177

[108] Diksha K, Bhavanam A, Giribabu D. Lignin and black liquor based composite hydrogels for the enhanced adsorption of malachite green dye from aqueous solutions: Kinetics, rheology and isotherm studies. International Journal of Biological Macromolecules. 2025;**296**:139613

[109] Bashir T, Dutta J, Masarat S, Rafatullah M, Asthana N, Khan AA, et al. Application of novel Chitosan-Lignin hydrogel beads for optimized removal of textile dye DB-218. Journal of Molecular Structure. 2025;**1324**:140892

[110] Roa K, Oyarce E, Boulett A, Pizarro G, Sánchez J. Lignin-based polymeric hydrogel materials for the removal of copper ions from water. Journal of Applied Polymer Science. 2025;**142**(9):e56536

[111] Li X, Li P, Chen W, Ren J, Wu W. Preparation and adsorption properties of lignin/cellulose hydrogel. Materials. 2023;**16**(12):4260

[112] Sheraz M, Sun X-F, Siddiqui A, Hu S, Song Z.Research Advances in Natural Polymers for Environmental Remediation. Polymers. 2025;**17**(5):559

Chapter 3

DNA-Based Hydrogel: Programmable Architectures and Application Frontiers

Daisuke Ishikawa

Abstract

DNA hydrogels represent a new generation of materials that combine DNA nanotechnology with materials science to create three-dimensional polymeric networks that are programmable and biocompatible. The architecture of their structure can be precisely managed through chemical or physical crosslinking and physical entanglement methods, which enable the creation of hydrogels with particular mechanical, chemical, and biological characteristics. Hydrogels show high sensitivity to multiple stimuli, including pH, temperature, enzymes, and biomolecules, making them suitable for drug delivery, biosensing, immunomodulation, tissue engineering, and catalysis applications. The application of aptamers, DNAzymes, and other nucleic acid motifs through functionalization methods leads to targeted therapeutic applications, sensitive detection methods, and dynamic cell growth scaffolding. The commercial manufacturing process faces two major obstacles: expensive production costs and biological environmental degradation of the material. Researchers are currently examining hybrid material methods to boost their durability and extend their possible applications in the medical and environmental fields. DNA hydrogels show how molecular programming enables the creation of advanced smart biomaterials that will revolutionize medical treatments and diagnostic tools in the future.

Keywords: DNA nanotechnology, DNA hydrogel, programmable biomaterial, stimuli-responsive hydrogel, chemical and physical crosslinking, DNA motif, drug delivery, biosensing, catalysis

1. Introduction

DNA nanotechnology has undergone remarkable development over the past 40 years, evolving from a theoretical concept into a multidisciplinary scientific field that integrates molecular biology, chemistry, physics, computer science, and materials engineering. The foundational work by Seeman [1] in the 1980s established DNA as not only a carrier of genetic information but also as a structural material capable of forming specific patterns via Watson–Crick base pairing. This breakthrough has enabled the design of DNA molecules with programmable structural functions. In 1998, Winfree et al. [2] provided experimental evidence for DNA tile self-assembly into well-ordered two-dimensional lattices, demonstrating the feasibility of using base-pairing rules to construct complex structures. The advent of DNA origami by Rothemund [3] in 2006 further expanded this field, allowing the creation of arbitrary nanoscale shapes from long single-stranded DNA scaffolds and hundreds of short staple strands. This milestone not only expanded the designability of structural DNA nanotechnology but also presented modular and shape-controllable building blocks that function as essential structural units for bridging nanoscale DNA architectures to macroscopic material assemblies. These advances have led to the construction of three-dimensional (3D) architectures [4] and modular DNA blocks with diverse geometries [5, 6]. This approach provided adjustable building units that connected DNA nanoscale components to create macroscopic structures, solving the historical challenge of moving from nanoscale to macroscale in structural DNA design [7, 8]. The modular frameworks developed into DNA hydrogels, which enabled researchers to convert nanoscale precision into complete 3D bulk structures. Dynamic DNA nanodevices have been developed to perform molecular computation, motion, and signal transduction [9–11], highlighting DNA's versatility as an active molecule.

Researchers have progressed from nanoscale breakthroughs to the development of macroscopic functional materials through programmable molecular interactions. DNA hydrogels are a leading example of the conversion of nanoscale programming into 3D bulk structures. DNA materials have shown rapid development in their structural and functional design; however, researchers have encountered multiple unresolved problems. The scalability of DNA synthesis, the mechanical fragility of soft networks, and nuclease degradation, which causes biological instability, remain major unresolved issues for DNA-based constructs, including hydrogels [12, 13]. This chapter focuses on essential knowledge gaps in DNA materials while presenting new solutions and applications to address these problems. The historical development and classification of DNA hydrogels are summarized in **Figure 1**.

The translation of these nanoscale programming capabilities into macroscopic materials has been a major area of research, resulting in the development of DNA hydrogels. These hydrogels are 3D polymer networks in which DNA acts as the primary or sole crosslinking agent [14, 15]. Integrating DNA into hydrogel matrices offers several advantages over conventional polymers, such as biocompatibility, high water content that mimics extracellular environments, and the ability to respond to stimuli via DNA strand hybridization and denaturation [16]. The first DNA hydrogel was created in 1996 by Nagahara and Matsuda [17], who formed a gel network via DNA hybridization with a synthetic polymer. Subsequently, hydrogels composed entirely of DNA were developed using enzymatic synthesis, demonstrating that DNA molecules can self-assemble into macroscopic gels without synthetic polymers [15]. Researchers have since expanded hydrogel design through physically crosslinked systems using dynamic junctions between complementary strands and chemically crosslinked systems using covalent bonds for network stabilization [18]. These diverse preparation methods provide a foundation for tailoring hydrogels for specific biomedical applications.

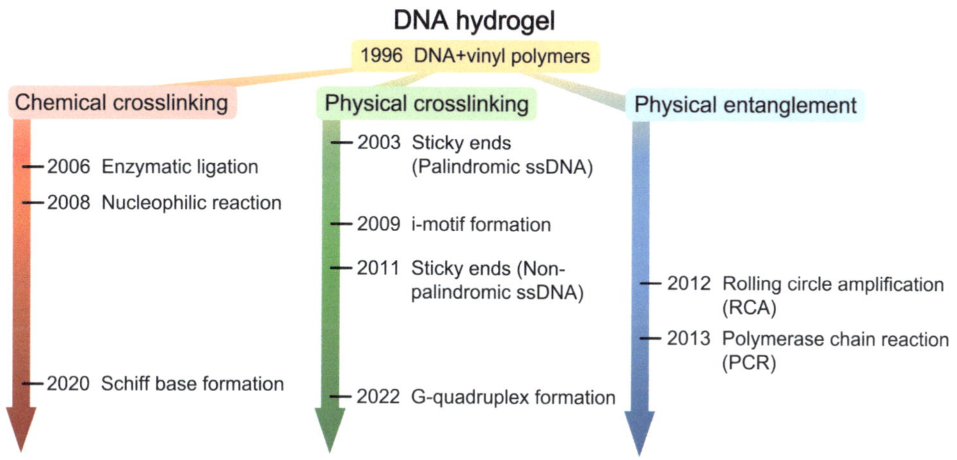

Figure 1.
Development roadmap of DNA hydrogels over the past few decades. They can be classified into three types based on their gel formation mechanisms: chemically crosslinked [15, 20, 21], physically crosslinked [22–25], and physically entangled [26, 27] hydrogels.

DNA hydrogels have emerged as powerful platforms for controlled drug delivery, biosensing, immunomodulation, and tissue engineering applications. Their multifunctionality is rooted in their programmable nature, biocompatibility, and structural similarity to natural extracellular matrices that support cell adhesion, proliferation, and differentiation [19]. The historical development of DNA hydrogels mirrors the evolution of DNA nanotechnology, from Seeman's structural work to the creation of complex, functional macroscopic materials. This interdisciplinary progress underscores the importance of collaboration among chemists, molecular biologists, polymer researchers, and engineers, and sets the stage for further exploration of DNA hydrogel classification, construction, and application.

2. Classification of DNA hydrogels

DNA hydrogels are generally classified into three categories based on themechanism of stabilization of their 3D network:

(1) Chemically crosslinked hydrogels. These compounds achieve stability through covalent bonds and coordination interactions, which create stable structural frameworks (**Figure 2a**).

(2) Physically crosslinked hydrogels. The Watson–Crick base-pairing or non-Watson–Crick base-pairing mechanisms function as a reversible and stimuli-sensitive network through non-covalent interactions (**Figure 2b**).

(3) Physically entangled hydrogels: Networks are generated through random topological interactions among extended DNA chains, unlike specific crosslinking reactions (**Figure 2c**).

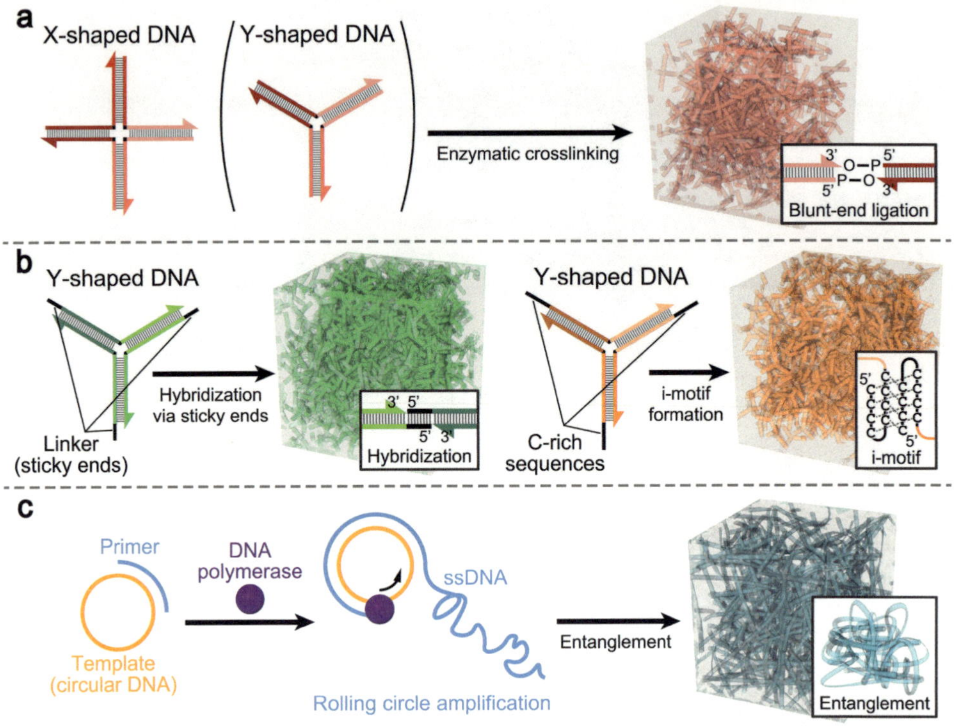

Figure 2.
Schematic representation of DNA hydrogel categories classified according to their stabilizing mechanisms. (a) Chemically crosslinked hydrogels formed by enzymatic ligation of X- or Y-shaped DNA motifs, generating stable networks through covalent bonds. (b) Physically crosslinked hydrogels assembled by non-covalent interactions, such as sticky-end hybridization of Y-shaped motifs or C-rich sequence-driven i-motif formation, enabling reversible and stimuli-responsive structures. (c) Physically entangled hydrogels produced from long single-stranded DNA generated by RCA of circular templates, in which the extended chains interweave to create topologically stabilized networks.

This classification system provides researchers with a conceptual framework for studying DNA hydrogel design principles, structural features, applications, benefits, and limitations [17, 28]. **Table 1** summarizes this classification. The following sections explain each category by showcasing the current developments of hybrid hydrogels that unite the properties of different classes.

2.1 Chemically crosslinked hydrogels

DNA hydrogels are stabilized through chemically crosslinked bonds, which are formed via covalent connections and coordination interactions between DNA strands modified with reactive chemical groups. These groups subsequently polymerize or couple with each other or with comonomers, producing strong, permanent, 3D networks that maintain their structure for extended periods and exhibit excellent mechanical properties for drug delivery, biosensing, and tissue engineering applications [28].

Enzyme-mediated strategies have also enabled the fabrication of DNA hydrogels entirely from branched DNA monomers that self-ligate through T4 DNA ligase under physiological conditions, yielding biocompatible and biodegradable networks capable

Classification	Formation mechanism	Characteristics	Refs.
Chemical crosslinking	Covalent bonds or coordination interactions between DNA strands (e.g., covalent coupling via amine or aldehyde functional groups; click chemistry).	Permanent, robust 3D networks; excellent mechanical strength and stability; tunable porosity and architecture; limited reversibility; less suitable for real-time sensing.	[15, 20, 21, 29]
Physical crosslinking	Noncovalent interactions via Watson–Crick base pairing or non-Watson–Crick base pairing, sequence-specific hybridization of sticky ends, cytosine-rich i-motifs, guanine-rich G-quadruplexes.	Reversible, stimuli-responsive (temperature, pH, enzymes, biomolecules); programmable assembly/disassembly; self-healing, biosensing, catalytic functions; lower mechanical strength than chemically crosslinked hydrogels.	[22, 23, 24, 25, 30, 31, 32, 33, 34]
Physical entanglement	Random topological entanglement of long DNA chains produced by polymerase chain reaction (PCR), rolling circle amplification (RCA), and enzymatic polymerization.	High water content, elasticity, and self-healing; simple preparation; mechanical properties inferior to chemically or physically crosslinked hydrogels; less controllable microstructure.	[26, 27, 35, 36, 37]

Table 1.
Classification of DNA hydrogels by stabilizing mechanism.

of encapsulating proteins or living cells with tunable porosity and shape fidelity (**Figure 3a**) [15]. In parallel, chemically crosslinked systems employing bifunctional epoxide reagents, such as ethylene glycol diglycidyl ether (EGDE), create covalent junctions between amino groups on nucleotide bases, forming stable gels with an elastic modulus exceeding 10^3 Pa and a nearly frequency-independent viscoelastic behavior (**Figure 3b**) [20]. Chemically crosslinked DNA hydrogels have evolved into functional materials that respond to external stimuli, such as ion coordination and molecular recognition. Polymeric networks with aptamer or DNAzyme motifs enable programmable sensing [28], whereas Ag^+-crosslinked systems exhibit reversible gel–sol transitions via cytosine–Ag^+–cytosine bridges [38].

2.2 Physically crosslinked hydrogels

DNA hydrogels that undergo physical crosslinking achieve Watson–Crick base pairing or non-Watson–Crick base pairing depending on the nucleotide sequence, which creates non-covalent interactions. The reversibility of DNA hybridization enables stimuli-responsive sol–gel transitions triggered by temperature [39], ionic strength [40], enzymatic activity [12], and target biomolecules [41]. These systems often employ branched DNA motifs, such as Y- and X-shaped structures with sticky ends, allowing precise control over gelation [42], network topology [43], and hierarchical structure [12], from nanoscale motifs to macroscopic, often porous [12], hydrogel networks [44]. The design of sticky-end sequences and motif geometry can further fine-tune the mechanical and responsive properties of the gel.

Physically crosslinked DNA hydrogels are formed by assembling DNA building blocks into 3D networks via non-covalent interactions. The most common crosslinking method is the hybridization of complementary sticky ends on branched DNA motifs, such as Y- or X-shaped structures, which create stable junctions and a robust hydrogel

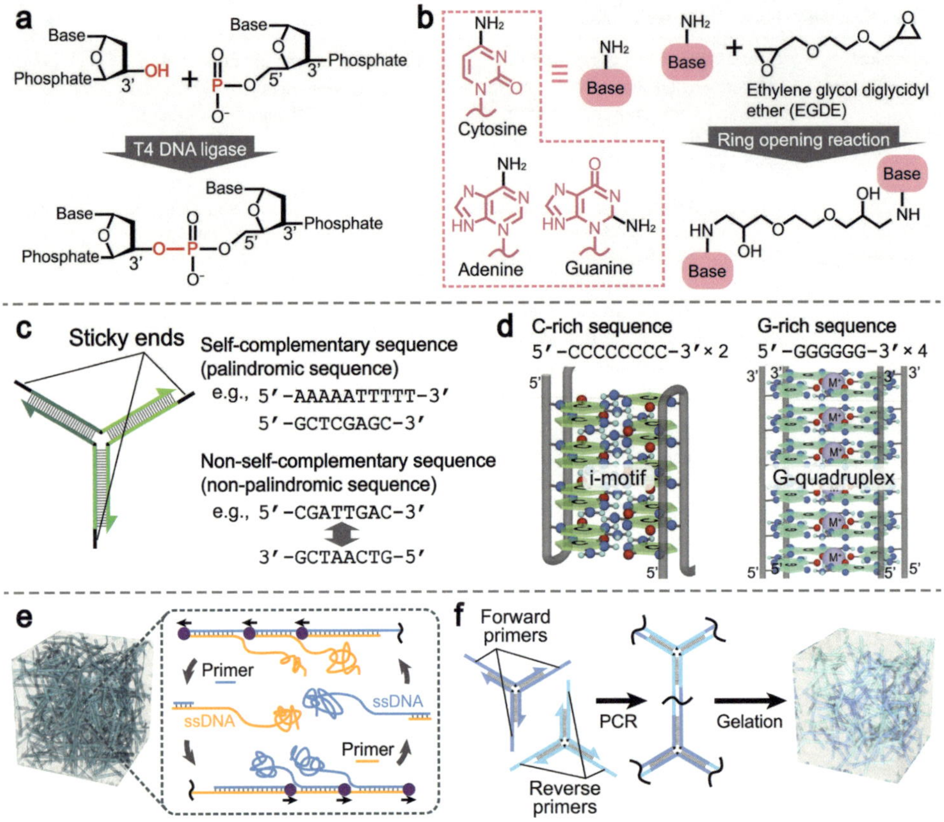

Figure 3.
Formation process of DNA hydrogels: (a) Enzyme-catalyzed ligation of blunt ends of DNA [15]. (b) Linking of nucleic acid bases via nucleophilic reaction of amino groups with diglycidyl ether [20]. (c) Crosslinking via hybridization of palindromic and non-palindromic sequences at sticky ends [23, 24, 33]. (d) Crosslinking via i-motif formation at C-rich ends [30] and G-quadruplex formation at G-rich ends [25]. (e) Physical entanglement formation through repeated single-stranded DNA generation by RCA [26]. (f) Physical entanglement formed by motif ends extended by PCR [27].

matrix (**Figure 3c**) [22–24, 30, 31]. In some cases, plasmids or linear DNA strands are crosslinked by short oligonucleotides that bridge multiple DNA chains, further stabilizing the network [31, 32]. Additionally, cytosine-rich sequences can fold into i-motif structures under mildly acidic conditions, forming intercalated $C \cdot C^+$ base pairs that provide reversible, pH-responsive crosslinks, enabling dynamic sol–gel transitions and structural tunability [30]. Guanine-rich sequences can self-assemble into four-stranded G-quadruplex structures that serve as additional crosslinking points in the hydrogel network (**Figure 3d**) [25]. These strategies enable hydrogel formation without covalent bonds, relying on the specificity and reversibility of DNA base pairing for precise and biocompatible network designs. The rational sequence design method enables programs to perform reversible assembly and disassembly operations through the detection of specific stimuli. Researchers have used enzyme-mediated assembly with temperature cycling [12] and dynamic DNA motifs [16] to develop hydrogels that exhibit adjustable porosity [12], self-healing properties [44], and multilevel responsiveness [28]. These DNA structures enable researchers to combine aptamers and DNAzymes as functional domains, which create biosensing [43] and catalytic capabilities [45]. Although these hydrogels exhibit excellent performance in dynamic and responsive applications, their mechanical properties remain inferior to those of chemically crosslinked systems.

2.3 Physically entangled hydrogels

The structural stability of physically entangled hydrogels stems from the generation of random DNA strands formed by RCA (**Figure 3e**) [26, 35] or PCR (**Figure 3f**) [27]. Physical crosslinking depends on specific base pairing through Watson–Crick and non-Watson–Crick interactions, but physical entanglement occurs through non-specific interchain contact between long DNA polymers. Topological stabilization emerges from DNA chains that randomly coil into each other without sequence-directed or covalent binding, creating an elastic network structure. Enzymatic methods of RCA and multi-primed chain amplification (MCA) [37, 46] enable the creation of hydrogels with high water content and self-healing properties through entangled systems. The DNA strands extend to form bulk gel networks that resemble synthetic polymer entanglements. The hierarchical structure develops through the combination of DNA molecular weight with chain entanglement density and macroscopic gel network formation, producing materials with high water content, elasticity, and self-healing properties [47].

The preparation of ultralong DNA strands through PCR, RCA, and enzymatic polymerization allows researchers to adjust the DNA concentration and chain length for network property control. The combination of polymers and nanoparticles with hydrogels leads to improved mechanical properties and additional functional capabilities [48]. The preparation process of these hydrogels is straightforward, but their mechanical properties are inferior to those of crosslinked systems, and their microstructure is less controllable.

2.4 Hybrid and composite hydrogels

Recent efforts have focused on developing hybrid and composite DNA hydrogels that combine the advantages of multiple classes. These materials integrate chemically and physically crosslinked networks [49, 50] or incorporate other polymers [8, 17, 51], nanoparticles [52–54], or bioactive molecules [55–57] to achieve enhanced mechanical strength, responsiveness, and multifunctionality. Hybrid hydrogels can be engineered to respond to multiple stimuli, support cell growth, and deliver therapeutic agents with high precision. The rational combination of different crosslinking strategies and material components expands the application potential of DNA hydrogels in advanced biomedical engineering and smart-material design. Hybrid DNA hydrogels show superior properties because their dual crosslinking approach surpasses the capabilities of single-component systems. The combination of chemical crosslinking for long-term stability and mechanical strength with physical crosslinking through cytosine-rich i-motif structures enables these systems to exhibit pH-responsive and reversible sol–gel transitions. The dual-network system demonstrates both shape-memory properties and controlled release functionality because the i-motif domains undergo structural changes between the folded and unfolded states when exposed to slightly acidic conditions [58]. The PEG chains that connect to oligonucleotides through covalent bonds establish a stable backbone framework that enables dynamic DNA hybridization and G-quadruplex formation to create reversible junctions for multi-stimuli responsiveness [59]. Creusen et al. [60] demonstrated how four-arm PEG-oligonucleotide conjugates assembled model network hydrogels that integrated covalent bonds for structural support with sequence-based hybridization for controlling viscoelastic properties and achieving reversible structural transformations. Hybrid structures unite permanent covalent frameworks with programmable DNA motifs to create hydrogel materials that dynamically execute biological functions while maintaining their structural integrity.

3. Application frontiers of DNA hydrogels

DNA hydrogels are 3D polymeric networks formed by the crosslinking of DNA strands, which leverage the unique programmability, molecular recognition, and biocompatibility of nucleic acids [16, 61]. Since Seeman's foundational work on DNA nanostructures [1], the field has rapidly evolved, integrating advances such as DNA origami [3], branched DNA motifs [12], and enzyme-catalyzed assembly [12]. Hydrogel properties are tailored through sequence design, crosslinking methods, and material hybridization, enabling them to detect multiple stimulus types, including pH changes [30, 62], temperature fluctuations [24], enzyme activity [63], light exposure [64, 65], and ionic concentration [66]. **Table 2** summarizes representative examples of these stimuli, their corresponding functions, and potential applications.

Stimuli	Triggers	Functions	Applications	Refs.
pH	Acidic/basic conditions	Gel–sol transition; volume change; shape-memory	pH-gated drug release; responsive matrices; information storage; soft actuators	[30, 62]
Metal ions	Pb^{2+}, Hg^{2+} and others	Gel–sol transition; visual detection	Heavy metal sensing; toxic ion sensing	[14, 74]
Enzymes	DNA ligase, restriction enzyme, DNAzyme	Gel–sol transition; controlled release	Drug/gene delivery; on-demand release; signal amplification	[15, 75]
Temperature	DNA duplex melting	Gel–sol transition	Responsive carriers; programmable scaffolds	[24]
Light	NIR/visible light, photo-activated self-assembly	Shape change; drug release; plasmonic response; light-induced bacterial inactivation.	Photolithography; photothermal therapy; dynamic optical materials; antimicrobial coatings	[64, 65]
Biomolecules	ATP, nucleic acids, proteins	Target recognition; gel–sol transition; fluorescence readout	Biosensing; logic gates; smart diagnostics	[41, 76]
Composite stimuli	NIR photothermal, Light, redox reactions	Dual-triggered gel–sol transition; shape-memory; self-healing	Synergistic cancer therapy; smart actuators	[39, 77]

This table shows the functions observed for the triggers and their potential applications.

Table 2.
Classification of DNA hydrogels based on major stimulus categories.

The incorporation of functional motifs, such as aptamers [59], DNAzymes [63], and i-motifs [67], enables DNA hydrogels to function as platforms for targeted drug delivery [68, 69], biosensing [59], immunomodulation [43, 67], regenerative medicine [61, 70, 71], smart materials [16, 23], and environmental remediation [71]. Recent research has also focused on improving the mechanical strength [70, 71], scalability, and integration of DNA hydrogels with nanomaterials [33, 72] or synthetic polymers [17, 73] to expand their utility in both biomedical and technological domains. This section provides a comprehensive analysis of the major application frontiers of DNA hydrogels, highlighting their distinct mechanisms, representative studies, and future directions.

3.1 Drug delivery

DNA hydrogels have transformed drug delivery systems through their ability to provide multiple release mechanisms that respond to environmental changes. The 3D DNA network structures of these hydrogels can protect various drugs, including small molecules, proteins, and nucleic acids, from enzymatic degradation and immune system attacks. For example, in these systems, protein encapsulation and stabilization against enzymatic degradation are achieved through reversible hybridization cross-linking [24]. On the other hand, nucleic acid protection is enabled by the formation of molecularly selective barriers that restrict nuclease access [71]. The drug-loading capacity of dendritic DNA hydrogels and RCA-based hydrogels is high, and their release mechanisms can be precisely controlled [68, 78]. Mo et al. [18] showed how DNA hydrogels can deliver gene editing and anticancer drugs through tumor microenvironment signal activation, which leads to targeted therapy with reduced side effects. DNA hydrogels containing cytosine-rich i-motifs or triple-stranded DNA regions break down in the acidic tumor environment to release chemotherapeutic agents in a controlled manner [30, 62, 69]. The application of DNA ligase, restriction enzymes, and DNA enzymes in enzymatically degradable hydrogels enables targeted drug release through specific enzymatic breakdown [15, 75]. The combination of nanoparticles with synthetic polymers in hybrid hydrogels leads to better drug delivery results and enables dual treatment approaches and simultaneous drug administration [28, 33, 72]. PEG-DNA hybrid networks unite the biocompatible nature of nucleic acids with the mechanical properties and manufacturing capabilities of synthetic polymers to create injectable self-healing drug delivery systems for sustained localized therapy [47, 59, 79]. The drug retention period and drug exposure time in tissues are prolonged because of these hybrid materials, leading to better therapeutic outcomes. Researchers have created new injectable self-healing hydrogels that form in situ for targeted treatments and gene-drug co-delivery systems that incorporate CRISPR/Cas9 gene-editing tools [21, 80]. The combination of pH and temperature sensitivity in dual-stimuli-responsive hydrogels produces unique release patterns, representing new approaches for medical precision [39, 77].

Challenges: The functional lifetime of DNA hydrogels and the drug release duration are limited by in vivo nuclease degradation, which reduces programmability [81].

Solutions: The development of phosphorothioate linkages and l-DNA strands with mirror chirality protects DNA from enzymatic degradation. The combination of PEG-DNA hybrid polymer frameworks and large-scale liquid-phase oligonucleotide synthesis provides both mechanical strength and production consistency to DNA-based materials [47, 60, 79].

DNA hydrogels have the potential to be next-generation drug delivery systems through the integration of sequence design with hybrid polymer frameworks and stimulus-controlled release mechanisms, which provide programmable and responsive systems with biological stability [68, 69, 81].

3.2 Biosensing

DNA hydrogels are essential biosensing materials because they transform molecular recognition events into measurable physical signals at the macroscopic scale. The 3D networks of DNA hydrogels function as biocompatible materials that can change their swelling state or collapse, as well as their optical properties, when they detect specific analytes [43]. The visual detection of metal ions, Pb^{2+} and Hg^{2+} [14, 74] is possible through DNAzyme-crosslinked hydrogels, while aptamer-functionalized hydrogels show high selectivity for proteins [82], small molecules [83], and pathogens [84].

The combination of microfluidic chips with point-of-care devices enables fast and quantitative multiplexed analyses of clinical and environmental samples [28, 85]. The system uses Boolean-logic-based hydrogel responses together with colorimetric or fluorescent readouts to achieve better sensitivity and user-friendly operation [86, 87]. DNA hydrogels have proven to be excellent biosensing matrices because they can be structurally programmed and exhibit reversible phase transitions. The immobilization of aptamers, DNAzymes, and other recognition motifs within crosslinked DNA networks enables selective responses to stimuli and target analytes, which produce measurable gel–sol transitions and optical changes [43, 45]. The base-pairing properties of DNA enable signal amplification through processes such as strand displacement, catalytic turnover, and hybridization cascades. The recognition of adenosine, ATP, thrombin, and heavy metal ions by aptamer-functionalized hydrogels leads to gel collapse and fluorescence recovery upon binding to their target molecules [43, 81]. The presence of Mg^{2+} or Pb^{2+} ions in DNAzyme-crosslinked hydrogels triggers strand cleavage, producing visible or electrochemical detection signals [45]. The pH-sensitive hydrogel properties enable it to function as a visual indicator for acidic environments, and enzyme-triggered sol-gel transitions create logic-gate biosensors that detect particular biochemical combinations [26, 55]. The combination of nucleic acid recognition with network responsiveness leads to the development of highly sensitive detection systems that operate without labels and can be reused. The combination of nanomaterials with microfabrication techniques enables the development of advanced sensing capabilities. The combination of DNA with polymers and nanoparticles creates hybrid materials that boost signal transmission through better electron transfer, enhanced optical detection, and mechanical reinforcement [47, 59]. The plasmonic color changes of Au-nanoparticle-embedded DNA hydrogels enable the visible detection of nucleic acids and pathogens [59, 82]. The integration of microfluidics into systems allows real-time continuous monitoring of small sample amounts, which creates potential applications for medical diagnostics and environmental monitoring.

Challenges: DNA hydrogels demonstrate excellent selectivity, but standard bulk gels exhibit poor mechanical strength and slow response times, which negatively affect the performance of miniaturized devices.

Solutions: The use of photolithography and 3D printing techniques to create microstructured thin-film DNA hydrogels reduces diffusion distances, enabling faster analyte detection [59]. The addition of conductive nanomaterials, including Au nanoparticles, graphene oxide, and carbon nanotubes, enhances both electronic signal transmission and mechanical durability of the system. The combination of responsive aptamers and DNAzymes in multiplexed arrays enables the fast, simultaneous detection of multiple analytes [45, 81].

DNA hydrogels function as advanced biosensing materials that merge specific molecular detection with structural changes and interface capabilities for optical and electronic systems, creating a link between nanoscale programming and large-scale analytical systems [43, 45, 59, 81].

3.3 Immunomodulation

DNA hydrogels show strong promise as immunomodulatory materials because they can control complex immune reactions through their programmable design and biocompatibility. Hydrogels become immunostimulatory through the integration of CpG motifs [55], peptide antigens [88], and cytokines [89], which activate dendritic cells, boost antigen presentation, and modify tumor environments. The use of injectable self-gelling DNA hydrogels as antigen depots enables an extended immune activation duration and enhanced vaccine effectiveness through the controlled release of immunostimulatory oligonucleotides [90]. Fan et al. [91] created tumor-sensitive DNA hydrogels, which release immune-checkpoint inhibitors and chemotherapeutic agents to enhance antitumor responses through controlled spatial and temporal delivery. DNA hydrogels offer distinct advantages over standard delivery systems because they can incorporate multiple immunomodulators within one matrix structure while creating a 3D space that promotes immune cell penetration and activation. These materials show great promise for cancer immunotherapy [92], vaccine development [90], allergy treatment, inflammation control, and wound healing applications [93, 94]. The excellent biocompatibility of DNA hydrogels, combined with their programmable degradation characteristics, allows for controlled immune system activation and clearance while minimizing side effects throughout the body. DNA hydrogel structures containing i-motif or G-quadruplex motifs are sensitive to acidic or oxidative tumor environments, enabling the targeted delivery of immunoadjuvants and checkpoint inhibitors to specific disease locations. The combination of DNA motifs with synthetic polymers or nanoparticles in hybrid systems leads to improved delivery stability and enhanced immunogenicity [47, 59, 79].

Challenges: DNA hydrogels effectively deliver antigens and immunostimulatory nucleic acids, but their therapeutic effects are limited by nuclease degradation and variable immune response patterns [81].

Solutions: The combination of phosphorothioate backbone modifications with PEG-DNA hybrid structures [81] creates stable structures that provide prolonged activation of the immune system. The combination of tumor-microenvironment-activatable designs [95] with cytokine [89] and antigen [88] co-delivery methods enables precise timing of immune response activation, which reduces unwanted inflammation and enhances treatment success.

DNA hydrogels with programmable properties serve as versatile systems for future immunotherapy development because they integrate specific molecular interactions with controlled drug delivery and biocompatible features to produce sustained immune system regulation [24, 47, 59].

3.4 Regenerative medicine

DNA hydrogels serve as adaptive frameworks in regenerative medicine, enabling cell attachment and growth, and directing tissue development through embedded biological signals. These materials show excellent potential for 3D cell culture [61], tissue engineering [96], and wound-healing [80] applications because they have adjustable mechanical properties and natural biocompatibility. Zhu et al. [19] showed that DNA hydrogels can duplicate the properties of the extracellular matrix properties to create tissue-like structures, which support stem-cell differentiation processes. Materials have evolved into smart scaffolds through the addition of growth factors, gene-editing tools, and peptide motifs, which enable them to detect local biochemical signals to direct cell fate during tissue repair processes [18, 97]. The combination of DNA–protein hybrid hydrogels with enzyme-catalyzed assemblies creates new functional capabilities that allow tissue integration and controlled structural modifications [45, 98]. The development of hydrogels that can dynamically connect with cells and their surrounding environment represents a fundamental requirement for creating advanced regenerative therapies in the future. DNA hydrogels have extensive biomedical applications through their use in bone and cartilage regeneration [96], neural tissue engineering [99], and extracellular vesicle-based tissue repair [95].

Challenges: DNA hydrogels show promise; however, they fail to achieve adequate mechanical strength and long-term stability in biological environments, which hinders their large-scale medical applications because DNA synthesis is complex and expensive [79, 100].

Solutions: The combination of DNA frameworks with synthetic polymers, such as PEG, and the use of nuclease-resistant l-DNA structures improve both the mechanical properties and the extended durability of these materials [100]. The combination of 3D printing methods with biomineralization-assisted assembly [101] and click chemistry-based DNA–peptide conjugation [97] enables the mass production of customized regenerative scaffolds.

DNA hydrogels serve as programmable, biocompatible platforms for regenerative medicine because they utilize molecular recognition, structural adaptability, and bioactivity to direct tissue reconstruction and healing processes [18, 19, 59, 96].

3.5 Smart materials and environmental applications

DNA hydrogels are at the forefront of smart materials and environmental science because they combine molecular programming with mechanical flexibility. The incorporation of G-quadruplexes [75], i-motifs, and triplex structures [62] within these hydrogels enables them to change between sol and gel states [62] while showing shape-memory properties [102] and self-healing abilities [103] through pH [62], temperature [64], and light [77] stimuli. The multiple functional responses of these materials make them suitable for soft robotic applications [8], information storage systems [104], and bioinspired actuators that replicate natural tissue mechanics [105]. DNA hydrogels serve as environmentally friendly adsorbents [71] and specific sensors for water cleaning and pollution cleanup operations in environmental applications [71, 106]. The DNAzyme-crosslinked networks in these materials enable the capture of toxic heavy-metal ions, such as Hg^{2+} and Pb^{2+}, while preserving their structural reversibility and recyclability [14, 74]. Das et al. [29] created dual-crosslinked DNA–nanoparticle hydrogels, which use visible light to kill bacteria, thus showing promise for antimicrobial coatings and wastewater treatment applications. The combination of nanozymes [107] with metal-organic frameworks (MOFs) [108] and quantum dots [109] expands the capabilities of DNA hydrogel systems for catalysis and sensing. The hybrid materials function as colorimetric or redox-based detectors and operate as photocatalytic or enzymatic reactors in mild aqueous solutions to support green chemistry and sustainable environmental monitoring. The development of flow-induced microfluidic assembly [110] and large-scale StarPEG-DNA network design [79] techniques has enhanced the mechanical properties and deployment readiness of DNA hydrogels for use in field applications.

Challenges: DNA hydrogels used in environmental applications and smart devices face three major barriers that prevent their widespread adoption: mechanical instability, slow response times, and expensive DNA synthesis costs [79].

Solutions: The combination of DNA frameworks with synthetic polymers [47, 59], nanoparticles [111], or MOF [108] inclusions strengthens their mechanical properties and enables faster responses and multiple applications. The combination of photo-patterning techniques [65] and microfluidic processing methods [110] enables the creation of smart actuators through precise structural design. The use of biomass-derived nucleic acids [29] provides both affordable and environmentally friendly solutions for environmental cleanup.

DNA hydrogels serve as intelligent, eco-responsive materials that combine biological adaptability with materials engineering capabilities to create new opportunities for soft robotics, environmental purification, and sustainable catalytic technologies [29, 59, 62].

3.6 Future perspectives and challenges

DNA hydrogels have evolved from theoretical nucleic-acid networks into functional, programmable materials that serve multiple biomedical, environmental, and catalytic purposes. The transition of DNA hydrogels from laboratory demonstrations to practical clinical and industrial applications faces several challenges. The development of DNA hydrogels requires researchers to combine mechanical strength with large-scale production methods and affordable DNA manufacturing techniques, while preserving their biological accuracy and molecular design capabilities [79, 110, 111]. Researchers have developed DNA origami modules, branched motifs, and polymeric backbones to create dynamic hierarchical structures [47, 59, 79]. The development of multiscale architectures will lead to the creation of self-healing constructs that can adapt to physiological stresses through mechanoadaptive responses. The combination of DNA hydrogels with synthetic biology and artificial-intelligence-driven design will boost the predictive modeling of structure–function relationships and create adaptive biomedical devices [112, 113]. The addition of photo-, thermal-, and enzyme-sensitive components to DNA hydrogels extends the range of stimulus-controlled systems [62, 64, 77]. DNA hydrogels can create complex tissue and catalytic interface patterns through the combination of microfluidic systems [110] and 3D printing technology [114]. The integration of nanoparticles [111] and MOFs [108] into biohybrid systems creates new possibilities for energy conversion, environmental cleanup, and soft robotic applications.

Challenge (1): Biostability and immunogenicity. The use of DNA hydrogels in vivo is restricted because enzymes break them down and the body produces unpredictable immune reactions [81].

Solution (1): The combination of phosphorothioate backbone modifications with PEG-DNA hybrid frameworks creates stable and biocompatible structures for extended periods of time [79, 81].

Challenge (2): Scalability and manufacturing costs. The production of pure oligonucleotides on an industrial scale requires high costs and significant energy consumption [101].

Solution (2): Enzymatic DNA synthesis methods, combined with biomass-based nucleic acid production, create sustainable and affordable manufacturing processes [29, 101].

Challenge (3): Integration and translation of data. Current DNA hydrogel systems need to be evaluated under real-life conditions, which differ from controlled laboratory environments.

Solution (3): The successful translation of DNA hydrogels to clinical and industrial applications depends on the creation of standardized biocompatibility tests, predictive mechanical models, and regulatory systems [112, 113].

DNA hydrogels will achieve their future potential through the combination of biological precision, materials engineering, and computational design to create soft matter systems that connect molecular and macroscopic scales [59, 112].

4. Catalytic applications of physically crosslinked DNA hydrogels

Physically crosslinked DNA hydrogels are promising materials for catalytic applications because they combine programmable properties with biocompatibility [37] and reversible network structures [67]. Hydrogels are developed through Watson–Crick base pairing, Y- or X-shaped DNA motif hybridization, and the use of dynamic nucleic acid structures, including i-motifs and triplexes [65, 112, 113]. The following sections investigate how these hydrogels function as catalyst supports through the catalytic activity of gold nanoparticles (Au NPs) and discuss additional catalyst support materials, including oxides, carbon-based scaffolds, and synthetic polymers.

A landmark study by Ishikawa et al. [33] provides a detailed demonstration of how the internal microenvironment of DNA hydrogels can be engineered to synthesize catalytic nanomaterials (**Figure 4a**). In their study, three Y-shaped DNA motifs (Y-18, Y-28, and Y-39), identical in sequence but differing in arm length, were rationally designed to systematically modulate the crosslinking density and internal porosity of hydrogels. These motifs spontaneously assemble via complementary sticky-end hybridization into physically crosslinked hydrogel particles without the need for chemical crosslinkers, forming well-defined 3D DNA networks. Immersion of the DNA hydrogel particles in an aqueous $HAuCl_4$ solution induced the electrostatic attraction of Au^{3+} ions to the negatively charged phosphate backbone of the DNA strands, followed by $NaBH_4$ reduction, which generated

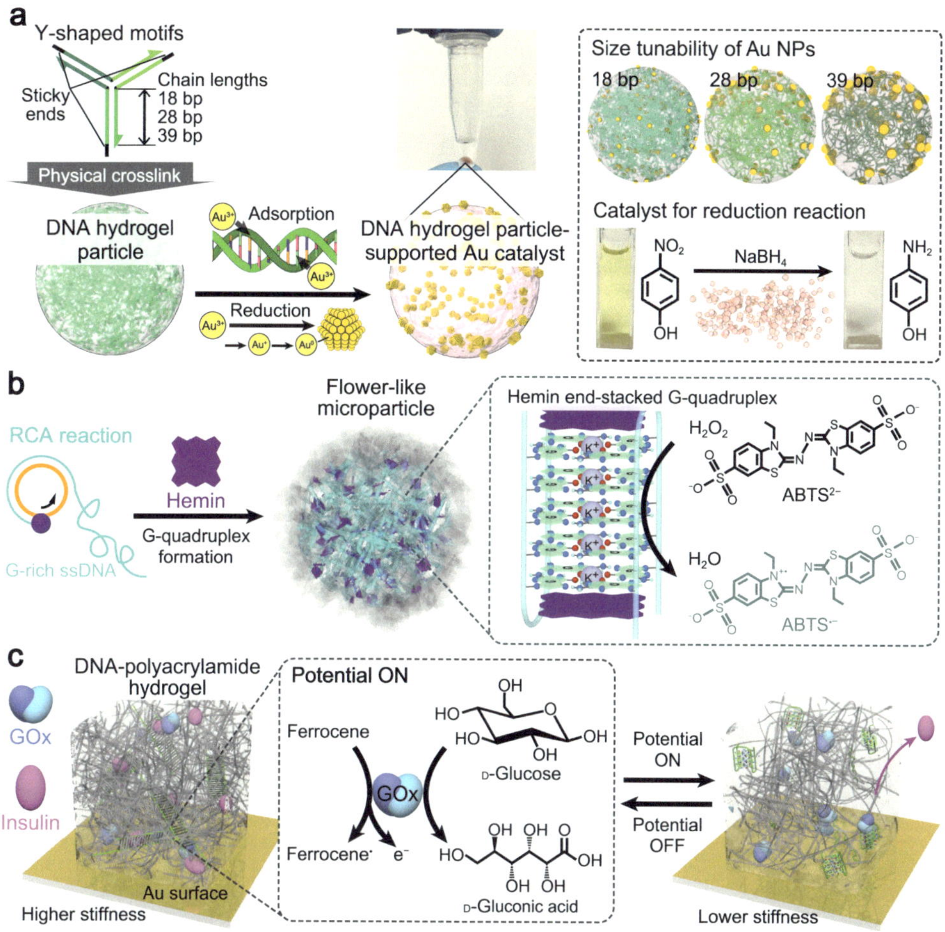

Figure 4.
(a) DNA hydrogel microparticles composed of Y-shaped motifs that function as tunable supports for Au NP catalysis [33]. (b) DNA hydrogel microparticles stabilized by a heme/G-quadruplex structure exhibiting stable horseradish peroxidase (HRP)-like catalytic activity [75]. (c) Stimuli-responsive glucose oxidase (GO_X)-loaded hydrogels with potential-induced switchable stiffness [117].

uniformly embedded Au NPs within and on the surface of the hydrogel matrix. Shorter motifs (Y-18) resulted in a denser hydrogel network that created small, uniform Au NPs with a diameter of 5.3 nm. The Y-39 motifs produced a less dense network, resulting in larger Au NPs measuring 12.2 nm in diameter. The size of the Au NPs formed within the DNA hydrogel particles was strongly dependent on the chain length of the Y-shaped DNA motifs constituting the hydrogel network. Shorter Y-shaped DNA motifs, such as Y-18, generated hydrogels with higher crosslinking densities and reduced internal pore sizes, as evidenced by the lower diffusion coefficients obtained from fluorescence recovery after photobleaching (FRAP) analysis. The FRAP results confirmed that the diffusivity within the network increased with motif length, demonstrating a direct relationship between DNA arm length, pore size, and local diffusion environment. In these dense networks, the diffusion and subsequent aggregation of Au atoms reduced from Au^{3+} ions were sterically hindered, thereby suppressing excessive crystal growth and resulting in the formation of small and uniform Au NPs. In contrast, hydrogels composed of longer Y-shaped DNA motifs (Y-28 and Y-39) exhibited lower crosslinking densities and larger internal voids, allowing the reduced Au atoms to migrate and coalesce freely. This facilitated the growth and fusion of Au nuclei within the more open hydrogel matrix, leading to the generation of larger and more polydisperse Au NPs, as observed in TEM images. These findings indicate that the size of the Au NPs is inversely correlated with the crosslinking density of the DNA hydrogel, which is governed by the length of the Y-shaped DNA motif arm. This structural tunability directly influenced catalytic behavior. In the 4-nitrophenol reduction reaction, the Y-18 hydrogel-supported Au NP catalyst exhibited approximately two-fold higher turnover frequency (TOF $\approx 4.3\ h^{-1}$) and four-fold higher apparent rate constant ($k \approx 8.9 \times 10^{-2}\ s^{-1}$) than those of the Y-39 system (TOF $\approx 2.0\ h^{-1}$, $k \approx 2.2 \times 10^{-2}\ s^{-1}$).

Thus, the structural designability of DNA motifs provides an effective means of regulating the spatial confinement of nanoparticle growth within hydrogel matrices. These quantitative findings clearly demonstrate that the nanoscale internal space and crosslinking density of DNA hydrogels significantly influence metal nucleation and active site accessibility. Programmable DNA hydrogels allow scientists to precisely manage nanoparticle development and catalytic behavior, providing an advantage over conventional materials [33, 115, 116].

From these and related studies, several general design principles for catalytic DNA hydrogels can be derived:

(1) Structural confinement governs nanoparticle formation and reactivity through control of crosslinking density and pore structure.

(2) Functional motif integration introduces biomimetic catalytic sites, such as G-quadruplexes and hemin-binding domains.

(3) Dynamic responsiveness enables reversible activation of catalysis via environmental triggers, such as pH or redox changes.

These strategies offer a unified approach to engineering catalytic DNA hydrogels at various length scales, from the molecular to the macroscopic level.

In parallel with the structural design principle demonstrated by Ishikawa et al. [33], researchers have explored functionalization approaches that incorporate catalytic domains or cofactors into the DNA framework. DNAzymes, aptamers, protein enzymes, and artificial cofactors, such as hemin, can be embedded into the DNA framework to achieve biomimetic or hybrid enzymatic catalysis [63]. Huang et al. [75] used RCA of G-rich templates to create a pure DNA hydrogel, which formed intermolecular G-quadruplexes that coordinated with hemin to produce peroxidase-mimicking catalytic centers (**Figure 4b**). The all-DNA hydrogel demonstrated superior thermal stability and maintained its catalytic activity for extended periods while catalyzing the oxidation of $ABTS^{2-}$ to $ABTS^{\bullet-}$ through H_2O_2 oxidation. The G-quadruplex/hemin motif demonstrates that nucleic acid structures can function as catalytic frameworks, enabling the construction of sustainable, metal-free oxidation systems.

The addition of stimuli-responsive catalytic systems further increases DNA hydrogel versatility. Fadeev et al. [117] reported a pH-responsive

DNA–polyacrylamide hybrid hydrogel crosslinked by duplex and i-motif strands, which exhibited reversible changes in stiffness and catalytic activity through biocatalytic acidification processes (**Figure 4c**). Glucose oxidase (GOx) loaded in the hydrogel triggered localized pH changes that switched the matrix between high- and low-stiffness states, thereby controlling both the rate of glucose oxidation and the release of coumarin-labeled insulin. Electrochemically induced localized acidification modulated the stiffness of i-motif crosslinked DNA hydrogels, thereby controlling the release of encapsulated insulin. This electrically controllable platform demonstrates a prototype of an autonomous bioelectronic "artificial pancreas," proving that DNA hydrogels can integrate biochemical sensing, energy conversion, and adaptive actuation within a single catalytic material.

Compared with conventional supports, DNA hydrogels offer unique advantages, including sequence programmability [7], molecular recognition [43], aqueous compatibility [67], and the creation of hydrated, biocompatible reaction spaces [118]. These properties enable mild and green catalytic processes and open pathways for biomedical, environmental, and energy-related applications. Zinchenko et al. [116] demonstrated that DNA hydrogels can template ultrasmall Au NPs under gentle aqueous conditions, suppressing aggregation and improving 4-nitrophenol reduction. This pioneering study exemplifies how DNA hydrogels provide environmentally benign catalytic microenvironments suitable for sustainable and biomedical applications. However, challenges remain regarding large-scale production [119] and long-term stability [120], particularly in biological environments where nuclease degradation can occur. Addressing these issues through molecular design and hybridization strategies will further expand the practical utility of DNA hydrogels as robust and sustainable catalytic platforms.

Research by Ishikawa et al. [33], Huang et al. [75], and Fadeev et al. [117] collectively demonstrates that programmable DNA hydrogel systems can simultaneously regulate nanoparticle assembly and molecular catalysis within a single soft-matter framework, providing a quantitative foundation for rational catalyst design through parameters such as turnover frequency, rate constant, and recyclability. Building on these insights, catalytic systems employing physically crosslinked DNA hydrogels as supports represent highly tunable and multifunctional platforms for the synthesis and application of catalytic nanomaterials. Their unique combination of programmable structures, biomimetic motifs, and dynamic responsiveness distinguishes them from conventional supports and positions them as promising candidates for next-generation catalytic systems in green chemistry, biosensing, and biomedical engineering.

5. Conclusions and future perspectives

DNA hydrogels are a new class of programmable biomaterials that can be designed as biocompatible and stimuli-responsive networks. Their distinctive features enable diverse applications, including drug delivery [68], biosensing [43], immunomodulation [91], tissue engineering [19], environmental remediation [71], and catalysis [33]. Through precise control of the DNA sequence and architecture, researchers can tune their mechanical [48], chemical [16], and biological [67] properties for both laboratory research and industrial applications.

In recent decades, DNA-polymer hybrids have evolved into highly programmable soft materials that combine nanoscale design with macroscopic functionality. The combination of their unique properties, including molecular recognition and water compatibility, makes DNA hydrogels suitable for multiple applications in biomedical and environmental fields. However, challenges remain in terms of cost-effective synthesis, long-term stability, and large-scale manufacturing of DNA hydrogels.

To accelerate their practical realization, future research goals should be prioritized based on feasibility and timelines:

(1) Short term (1–3 years): Development of high-throughput and low-cost DNA synthesis and amplification technologies [60] to reduce production costs and facilitate broader applications.

(2) Mid-term (3–5 years): Engineering hybrid DNA–polymer hydrogels that combine synthetic polymers to enhance mechanical robustness and resistance to enzymatic degradation [51]. These systems should balance structural stability and dynamic responsiveness for in vivo use.

(3) Long term (5–10 years): Implement microbial and cell-free production routes to enable the sustainable and scalable manufacturing of DNA hydrogels [119] for industrial-level translation and environmental deployment.

Achieving these goals requires interdisciplinary collaboration across different fields. Synergistic efforts between material scientists and process engineers can refine RCA and enzymatic polymerization techniques, allowing for the efficient and large-scale production of DNA hydrogels. Close cooperation between bioengineers and immunologists has led to the development of hydrogels designed for cancer immunotherapy and precise immune modulation.

The translational roadmap functions as a bridge that enables laboratory discoveries to transition into medical treatment applications. The first requirement is the use of multiple large animal models for the complete validation of biocompatibility and biodegradability to establish safety and biological stability. The second phase of development focuses on creating Phase I clinical trials for localized drug and gene delivery systems, which focus on precise dosage control and enhanced targeting accuracy. The third phase requires the development of complete regulatory systems and testing methods for DNA-based biomaterials to achieve reproducible results and meet ethical standards. The combination of programmable chemistry with bioengineering and clinical translation will enable the development of DNA hydrogels from experimental tools into practical therapeutic systems for precise medical treatments and enduring biotechnological applications.

Glossary of key terms

Aptamer
A short single-stranded DNA or RNA molecule, selected through Systematic Evolution of Ligands by Exponential Enrichment (SELEX), binds with high affinity and specificity to a target molecule, such as a protein, ion, or small molecule. Aptamers impart molecular recognition and biosensing functions to DNA hydrogels and are widely used as recognition modules in stimuli-responsive and therapeutic hydrogel systems.

Chemically crosslinked hydrogel
A hydrogel stabilized by covalent or coordination bonds between DNA strands or between DNA and comonomers. These networks are generally mechanically strong and stable, but some chemical crosslinks can be reversible, depending on the chemistry used (e.g., imine/Schiff base or disulfide bonds). Such frameworks are suitable for long-term or robust biomedical applications.

CRISPR/Cas9
A gene-editing tool that can be encapsulated or delivered via DNA hydrogels to enhance protection and localized release for genome engineering therapies. The use of DNA hydrogels for CRISPR/Cas9 delivery is a promising, but still emerging, application.

DNA hydrogel
A three-dimensional polymeric network in which DNA strands act as crosslinkers. They exhibit biocompatibility, programmability, and responsiveness to stimuli such as pH, light, temperature, enzymes, and ions. Depending on the crosslinking mechanism, DNA hydrogels are categorized as chemically, physically, or hybrid crosslinked systems.

DNAzyme
A catalytic DNA molecule (deoxyribozyme) that can catalyze specific chemical reactions (e.g., metal-ion-dependent cleavage). The incorporation of DNAzymes confers enzyme-like or signal-amplifying properties to hydrogels. Hemin–G-quadruplex DNAzymes are commonly used for peroxidase-mimicking bioelectrocatalysis.

G-quadruplex (G4)
A guanine-rich, four-stranded DNA structure stabilized by Hoogsteen hydrogen bonding and monovalent cations (K^+, Na^+), frequently used as a dynamic crosslinking motif or catalytic site (e.g., hemin–G4 DNAzyme). Its reversible folding provides tunable structural and redox properties in DNA-based nanomaterials.

G-quadruplex/hemin catalytic center
A hybrid catalytic site formed by a hemin (Fe-protoporphyrin IX) molecule bound to a G-quadruplex DNA mimics peroxidase activity in catalytic DNA hydrogels. This motif enables electron-transfer reactions and bioelectrocatalytic sensing interfaces.

Hybrid (composite) DNA hydrogel
A multi-component material combining DNA networks with synthetic polymers, nanoparticles, proteins, peptides, or inorganic materials provides enhanced mechanical strength and multifunctionality. For example, PEG-DNA or AuNP-DNA composites offer improved toughness, conductivity, or catalytic capacity.

i-motif

A cytosine-rich DNA secondary structure that folds into an intercalated tetraplex under acidic pH forms reversible $C{\cdot}C^+$ base pairs and enables pH-responsive switching in DNA hydrogels and nanodevices. It is often paired with G-quadruplex motifs to design dual pH-responsive systems that mimic cellular microenvironments.

Non-Watson–Crick base pairing

Non-canonical interactions, such as Hoogsteen, wobble, and mismatched base pairs, provide structural flexibility and alternative recognition motifs in DNA assemblies, expanding the design possibilities beyond standard duplex hybridization.

PEG-DNA hybrid hydrogel

A composite hydrogel of polyethylene glycol (PEG) and DNA combines the mechanical robustness of polymers with the biocompatibility and programmability of DNA. PEG crosslinking suppresses enzymatic degradation while maintaining hydration and diffusivity.

Physically crosslinked hydrogel

A hydrogel is assembled by non-covalent interactions such as base pairing, sticky-end hybridization, or i-motif/G-quadruplex formation. These hydrogels are reversible and stimuli-responsive but have lower mechanical strength. Their reversibility enables self-healing and shape-memory functions under environmental control.

Physically entangled hydrogel

A network formed by the random entanglement (not crosslinking) of long DNA chains, typically produced by RCA or PCR. These hydrogels are elastic, have high water content, and exhibit self-healing behavior due to physical entanglement rather than chemical or base-pair crosslinking.

Polymerase Chain Reaction

A thermal cycling process for amplifying DNA is also used to produce long DNA chains for physically entangled hydrogels. PCR-based hydrogels enable sequence-controlled architecture at the macroscopic level.

Rolling Circle Amplification

An enzymatic DNA amplification method that produces long single-stranded DNA from a circular template, generating entangled or crosslinkable networks for hydrogel formation. Frequently used for in situ hydrogel growth on surfaces or within living systems.

Self-healing

The ability of a hydrogel to repair its internal structure spontaneously after mechanical damage through reversible DNA interactions is enabled by hybridization-mediated strand exchange or dynamic hydrogen bonding.

Shape-memory

A property that allows a hydrogel to recover its original shape after deformation upon specific triggers such as pH or temperature enabled by reversible crosslinking motifs or supramolecular interactions within the network.

Stimuli-responsive hydrogel

A DNA hydrogel that changes its state or properties in response to external stimuli, such as pH, temperature, light, or biomolecules, enables controlled release or sensing. Representative systems include photo-switchable azobenzene–DNA, redox-responsive disulfide-linked gels, and enzyme-degradable matrices.

Watson–Crick base pairing

The canonical hydrogen-bonding pattern between adenine–thymine (A–T) and guanine–cytosine (G–C) bases forms the foundation of DNA duplex and hybridization processes.

Y-shaped DNA motif

A branched DNA structure with three arms is used as a programmable building block to control the crosslinking density and internal porosity of DNA hydrogels. The arm length dictates pore size, diffusivity, and nanoparticle templating capability within DNA hydrogels.

Acknowledgments

This study was supported by a Grant-in-Aid for Transformative Research Areas (A) "Materials Science of Meso-Hierarchy" (24H01702) from MEXT; JSPS KAKENHI (24K22332); PRESTO "Pioneering the Synergy of Information and Physics Connected by Optics and Photonics" (JPMJPR24L2) of JST; and the Research Center for Biomedical Engineering.

The author acknowledges the use of Paperpal for language polishing of the manuscript.

Conflict of Interest

The author declares no conflict of interest.

Author details

Daisuke Ishikawa
Laboratory for Biomaterials and Bioengineering, Institute of Integrated Research, Institute of Science Tokyo, Tokyo, Japan

*Address all correspondence to: ishikawa.daisuke@tmd.ac.jp, ishikawa.d.aa70@m.isct.ac.jp

References

[1] Seeman NC. Nucleic-acid junctions and lattices. Journal of Theoretical Biology. 1982;**99**(2):237–247. DOI: 10.1016/0022-5193(82)90002-9

[2] Winfree E, Liu FR, Wenzler LA, Seeman NC. Design and self-assembly of two-dimensional DNA crystals. Nature. 1998;**394**(6693):539–544. DOI: 10.1038/28998

[3] Rothemund PWK. Folding DNA to create nanoscale shapes and patterns. Nature. 2006;**440**(7082):297–302. DOI: 10.1038/nature04586

[4] Douglas SM, Dietz H, Liedl T, Hogberg B, Graf F, Shih WM. Self-assembly of DNA into nanoscale three-dimensional shapes. Nature. 2009;**459**(7245):414–8. DOI: 10.1038/nature08016

[5] Ke YG, Ong LL, Shih WM, Yin P. Three-Dimensional structures self-assembled from DNA bricks. Science. 2012;**338**(6111):1177–1183. DOI: 10.1126/science.1227268

[6] Ke YG, Ong LL, Sun W, Song J, Dong MD, Shih WM, et al. DNA brick crystals with prescribed depths. Nature Chemistry. 2014;**6**(11):994–1002. DOI: 10.1038/nchem.2083

[7] Sontakke VA, Yokobayashi Y. Programmable macroscopic self-assembly of DNA-Decorated hydrogels. Journal of the American Chemical Society. 2022;**144**(5): 2149–2155. DOI: 10.1021/jacs.1c10308

[8] Zhao Z, Wang C, Yan H, Liu Y. Soft robotics programmed with double crosslinking DNA hydrogels. Advanced Functional Materials. 2019;**29** (45):1905911. DOI: 10.1002/adfm.201905911

[9] Chatterjee G, Dalchau N, Muscat RA, Phillips A, Seelig G. A spatially localized architecture for fast and modular DNA computing. Nature Nanotechnology. 2017;**12**(9):920–927. DOI: 10.1038/nnano.2017.127

[10] Siti W, Too HL, Anderson T, Liu XR, Loh IY, Wang ZS. Autonomous DNA molecular motor tailor-designed to navigate DNA origami surface for fast complex motion and advanced nanorobotics. Science Advances. 2023;**9** (38):eadi8444. DOI: 10.1126/sciadv.adi8444

[11] Ishikawa D, Suzuki Y, Kurokawa C, Ohara M, Tsuchiya M, Morita M, et al. DNA origami nanoplate-based emulsion with nanopore function. Angewandte Chemie International Edition. 2019;**58** (43):15299–15303. DOI: 10.1002/anie.201908392

[12] Du XX B, YH HPP, Wang CY, Guo WW. Hierarchically structured DNA-Based hydrogels exhibiting enhanced enzyme-responsive and mechanical properties. Advanced Functional Materials. 2020;**30** (51):2006305. DOI: 10.1002/adfm.202006305

[13] Li YJ, Chen RF, Zhou BN, Dong YC, Liu DS. Rational design of DNA hydrogels based on molecular dynamics of polymers. Advanced Materials (Deerfield Beach, Fla.). 2024;**36**:2307129. DOI: 10.1002/adma.202307129

[14] Lin HX, Zou Y, Huang YS, Chen J, Zhang WY, Zhuang ZX, et al. DNAzyme

crosslinked hydrogel: A new platform for visual detection of metal ions. Chemical Communications. 2011;**47**(33):9312–9314. DOI: 10.1039/c1cc12290h

[15] Um SH, Lee JB, Park N, Kwon SY, Umbach CC, Luo D. Enzyme-catalysed assembly of DNA hydrogel. Nature Materials. 2006;**5**(10):797–801. DOI: 10.1038/nmat1741

[16] Kahn JS, Hu YW, Willner I. Stimuli-Responsive DNA-Based hydrogels: From basic principles to applications. Accounts of Chemical Research. 2017;**50**(4):680–690. DOI: 10.1021/acs.accounts.6b00542

[17] Nagahara S, Matsuda T. Hydrogel formation via hybridization of oligonucleotides derivatized in water-soluble vinyl polymers. Polymer Gels and Networks. 1996;**4**(2):111–127. DOI: 10.1016/0966-7822(96)00001-9

[18] Mo FL, Jiang K, Zhao D, Wang YQ, Song J, Tan WH. DNA hydrogel-based gene editing and drug delivery systems. Advanced Drug Delivery Reviews. 2021;**168**:79–98. DOI: 10.1016/j.addr.2020.07.018

[19] Zhu ZY, Yang YM, Jiang Y, Gu TY, Siow L, Gao YX, et al. DNA hydrogels in tissue engineering: From molecular design to next-generation biomedical applications. Advanced Healthcare Materials. 2025;**14**(13):2500192. DOI: 10.1002/adhm.202500192

[20] Topuz F, Okay O. Rheological behavior of responsive DNA hydrogels. Macromolecules. 2008;**41**(22):8847–8854. DOI: 10.1021/ma801414p

[21] Basu S, Pacelli S, Paul A. Self-healing DNA-based injectable hydrogels with reversible covalent linkages for controlled drug delivery. Acta Biomaterialia. 2020;**105**:159–169. DOI: 10.1016/j.actbio.2020.01.021

[22] Matsuura K, Yamashita T, Igami Y, Kimizuka N. 'Nucleo-nanocages': Designed ternary oligodeoxyribonucleotides spontaneously form nanosized DNA cages. Chemical Communications. 2003;**3**:376–377. DOI: 10.1039/b210139d

[23] Matsuura K, Masumoto K, Igami Y, Fujioka T, Kimizuka N. In situ observation of spherical DNA assembly in water and the controlled release of bound dyes. Biomacromolecules. 2007;**8**(9):2726–2732. DOI: 10.1021/bm070357w

[24] Xing YZ, Cheng EJ, Yang Y, Chen P, Zhang T, Sun YW, et al. Self-Assembled DNA hydrogels with designable thermal and enzymatic responsiveness. Advanced Materials. 2011;**23**(9):1117–1121. DOI: 10.1002/adma.201003343

[25] Fabrini G, Minard A, Brady RA, Di Antonio M, Di Michele L. Cation-Responsive and photocleavable hydrogels from noncanonical amphiphilic DNA nanostructures. Nano Letters. 2022;**22**(2):602–611. DOI: 10.1021/acs.nanolett.1c03314

[26] Lee J, Peng SM, Yang DY, Roh YH, Funabashi H, Park N, et al. A mechanical metamaterial made from a DNA hydrogel. Nature Nanotechnology. 2012;7(12):816–820. DOI: 10.1038/nnano.2012.211

[27] Hartman MR, Yang DY, Tran TNN, Lee K, Kahn JS, Kiatwuthinon P, et al. Thermostable branched DNA nanostructures as modular primers for polymerase chain reaction. Angewandte

Chemie International Edition. 2013;**52**(33):8699–8702. DOI: 10.1002/anie.201302175

[28] Li F, Tang JP, Geng JH, Luo D, Yang DY. Polymeric DNA hydrogel: Design, synthesis and applications. Progress in Polymer Science. 2019;**98**:101163. DOI: 10.1016/j.progpolymsci.2019.101163

[29] Das G, Nayak S, Kotness DK, Das P. A biomass-derived dual crosslinked DNA-nanoparticle hydrogel for visible light-induced photodynamic bacterial inactivation. Soft Matter. 2023;**19**(48):9511–9519. DOI: 10.1039/d3sm01400b

[30] Cheng EJ, Xing YZ, Chen P, Yang Y, Sun YW, Zhou DJ, et al. A pH-Triggered, fast-responding DNA hydrogel. Angewandte Chemie International Edition. 2009;**48**(41):7660–7663. DOI: 10.1002/anie.200902538

[31] Nishida Y, Ohtsuki S, Araie Y, Umeki Y, Endo M, Emura T, et al. Self-assembling DNA hydrogel-based delivery of immunoinhibitory nucleic acids to immune cells. Nanomedicine: Nanotechnology, Biology and Medicine. 2016;**12**(1):123–130. DOI: 10.1016/j.nano.2015.08.004

[32] Nöll T, Wenderhold-Reeb S, Schönherr H, Nöll G. Pristine DNA hydrogels from biotechnologically derived plasmid DNA. Angewandte Chemie International Edition. 2017;**56**(39):12004–12008. DOI: 10.1002/anie.201705001

[33] Ishikawa D, Maruyama K, Hara M. DNA motifs hydrogel microparticles as gold nanocatalyst support material. Particle & Particle Systems Characterization. 2025;**42**(2):2400060. DOI: 10.1002/ppsc.202400060

[34] Chai HX, Kayanuma K, Suzuki H, Takinoue M. DNA condensates via entanglement of string-like structures based on anisotropic nanotetrahedra. JACS Au. 2025;**5**(7):3249–3261. DOI: 10.1021/jacsau.5c00421

[35] Yao C, Zhang R, Tang JP, Yang DY. Rolling circle amplification (RCA)-based DNA hydrogel. Nature Protocols. 2021;**16**(12):5460–5483. DOI: 10.1038/s41596-021-00621-2

[36] Xu L, Wang H, Yang YN, Zhang H, Fang SQ, Zhao YC, et al. A straightforward process manipulates the dramatic morphological changes of DNA rolling circle amplification products. Nanoscale. 2025;**17**(5):2444–2450. DOI: 10.1039/d4nr04501g

[37] Hanif W, Yadav I, Hasan E, Alsulaiman D. Programmable all-DNA hydrogels based on rolling multi-primed chain amplification products. APL Bioengineering. 2023;**7**(4):046106. DOI: 10.1063/5.0169063

[38] Guo WW, Qi XJ, Orbach R, Lu CH, Freage L, Mironi-Harpaz I, et al. Reversible Ag^{+}-crosslinked DNA hydrogels. Chemical Communications. 2014;**50**:4065–4068

[39] He PP, Du XX, Cheng Y, Gao Q, Liu C, Wang XW, et al. Thermal-Responsive MXene-DNA Hydrogel for near-infrared light triggered localized photothermal-chemo synergistic cancer therapy. Small. 2022;**18**(40):2200263. DOI: 10.1002/smll.202200263

[40] Fern J, Shi RH, Liu YX, Xiong Y, Gracias DH, Schulman R. Swelling characteristics of DNA polymerization

gels. Soft Matter. 2023;**19**(34): 6525–6534. DOI: 10.1039/d3sm00321c

[41] Oishi M, Nakatani K. Dynamically programmed switchable DNA hydrogels based on a DNA circuit mechanism. Small. 2019;**15**(15):1900490. DOI: 10.1002/smll.201900490

[42] Yang B, Zhao ZH, Pan YF, Xie JY, Zhou BN, Li YJ, et al. Shear-Thinning and designable responsive supramolecular DNA hydrogels based on chemically branched DNA. ACS Applied Materials & Interfaces. 2021;**13** (41):48414–48422. DOI: 10.1021/acsami.1c15494

[43] Xi LQ, Shang YT, ZZ W, Wang J, Wu QP, Shen YZ, et al. Programmable DNA hydrogels for biosensing and point-of-care test. Coordination Chemistry Reviews. 2024;**518**:216084. DOI: 10.1016/j.ccr.2024.216084

[44] Wang C, Zhang JJ. Recent advances in stimuli-responsive DNA-based hydrogels. ACS Applied Bio Materials. 2022;**5**(5):1934–1953. DOI: 10.1021/acsabm.1c01197

[45] Xiang BB, He KY, Zhu R, Liu ZL, Zeng S, Huang Y, et al. Self-assembled DNA hydrogel based on enzymatically polymerized DNA for protein encapsulation and enzyme/DNAzyme hybrid cascade reaction. ACS Applied Materials & Interfaces. 2016;**8** (35):22801–22807. DOI: 10.1021/acsami.6b03572

[46] Stoll H, Steinle H, Stang K, Kunnakattu S, Scheideler L, Neumann B, et al. Generation of Large-Scale DNA Hydrogels with Excellent Blood and Cell Compatibility. Macromolecular Bioscience. 2017;**17** (4):1600252. DOI: 10.1002/mabi.201600252

[47] Uzumcu AT, Guney O, Okay O. Highly stretchable DNA/Clay hydrogels with self-healing ability. ACS Applied Materials & Interfaces. 2018;**10** (9):8296–8306. DOI: 10.1021/acsami.8b00168

[48] Cao DJ, Xie YJ, Song J. DNA hydrogels in the perspective of mechanical properties. Macromolecular Rapid Communications. 2022;**43** (19):2200281. DOI: 10.1002/marc.202200281

[49] Li C, Rowland MJ, Shao Y, Cao TY, Chen C, Jia HY, et al. Responsive double network hydrogels of interpenetrating DNA and CB[8] host-guest supramolecular systems. Advanced Materials. 2015;**27**:3298–3304

[50] Cao TY, Jia HY, Dong YC, Gui SB, Liu DS. In situ formation of covalent second network in a DNA supramolecular hydrogel and its application for 3D cell imaging. ACS Applied Materials & Interfaces. 2020;**12** (4):4185–4192. DOI: 10.1021/acsami.9b11534

[51] Li C, Chen P, Shao Y, Zhou X, Wu YZ, Yang ZQ, et al. A writable polypeptide-DNA hydrogel with rationally designed multi-modification sites. Small. 2015;**11**(9–10):1138–1143. DOI: 10.1002/smll.201401906

[52] Fukunaka Y, Iwanaga K, Morimoto K, Kakemi M, Tabata Y. Controlled release of plasmid DNA from cationized gelatin hydrogels based on hydrogel degradation. Journal of Controlled Release. 2002;**80** (1–3):333–343. DOI: 10.1016/s0168-3659(02)00026-3

[53] Wang SQ, Liu YH, Sun QW, Zeng BW, Liu C, Gong LM, et al. Triple cross-linked dynamic responsive

hydrogel loaded with selenium nanoparticles for modulating the inflammatory microenvironment via PI3K/Akt/NF-κB and MAPK signaling pathways. Advanced Science. 2023; **10**(31). DOI: 10.1002/advs.202303167

[54] Chen YF, Chiou YH, Chen YC, Jiang YS, Lee TY, Jan JS. ZnO-loaded DNA nanogels as neutrophil extracellular trap-like structures in the treatment of mouse peritonitis. Materials Science and Engineering: C. 2021;131. DOI: 10.1016/j.msec.2021.112484

[55] Nishikawa M, Ogawa K, Umeki Y, Mohri K, Kawasaki Y, Watanabe H, et al. Injectable, self-gelling, biodegradable, and immunomodulatory DNA hydrogel for antigen delivery. Journal of Controlled Release. 2014;**180**:25–32

[56] Zhang J, Guo YY, Pan GF, Wang P, Li YH, Zhu XY, et al. Injectable drug-conjugated DNA hydrogel for local chemotherapy to prevent tumor recurrence. ACS Applied Materials & Interfaces. 2020;**12**(19):21441–21449. DOI: 10.1021/acsami.0c03360

[57] Phan VHG, Duong HTT, Thambi T, Nguyen TL, Turabee MH, Yin Y, et al. Modularly engineered injectable hybrid hydrogels based on protein-polymer network as potent immunologic adjuvant. Biomaterials. 2019;**195**:100–110

[58] Hu YW, Lu CH, Guo WW, Aleman-Garcia MA, Ren JT, Willner I. A shape memory acrylamide/DNA hydrogel exhibiting switchable dual pH-responsiveness. Advanced Functional Materials. 2015;**25**(44):6867–6874. DOI: 10.1002/adfm.201503134

[59] Li J, Mo LT, Lu CH, Fu T, Yang HH, Tan WH. Functional nucleic acid-based hydrogels for bioanalytical and biomedical applications. Chemical Society Reviews. 2016;**45**(5):1410–1431. DOI: 10.1039/c5cs00586h

[60] Creusen G, Akintayo CO, Schumann K, Walther A. Scalable one-pot-liquid-phase oligonucleotide synthesis for model network hydrogels. Journal of the American Chemical Society. 2020;**142**(39):16610–16621. DOI: 10.1021/jacs.0c05488

[61] Chen C, Zhou JY, Chen J, Liu HJ. Advances in DNA supramolecular hydrogels for tissue engineering. Macromolecular Bioscience. 2022;**22**(12):2200152. DOI: 10.1002/mabi.202200152

[62] Hu YW, Ke YJ, Willner I. A pH-Cascaded DNA hydrogel mediated by reconfigurable A-motif duplex, i-Motif quadruplex, and T•A-T triplex structures. Advanced Functional Materials. 2023;**33**(45):2304966. DOI: 10.1002/adfm.202304966

[63] Ma YZ, Duan XC, Huang JY. DNA hydrogels as functional materials and their biomedical applications. Advanced Functional Materials. 2024;**34**(3):2309070. DOI: 10.1002/adfm.202309070

[64] Wang C, Liu X, Wulf V, Vazquez-Gonzalez M, Fadeev M, Willner I. DNA-Based hydrogels loaded with Au nanoparticles or Au nanorods: thermoresponsive plasmonic matrices for shape-memory, self-healing, controlled release, and mechanical applications. ACS Nano. 2019;**13**(3):3424–3433. DOI: 10.1021/acsnano.8b09470

[65] Kasahara Y, Sato Y, Masukawa MK, Okuda Y, Takinoue M. Photolithographic shape control of DNA

hydrogels by photo-activated self-assembly of DNA nanostructures. APL Bioengineering. 2020;**4**(1):016109. DOI: 10.1063/1.5132929

[66] Wu YF, Wang DY, Willner I, Tian Y, Jiang L. Smart DNA hydrogel integrated nanochannels with high ion flux and adjustable selective ionic transport. Angewandte Chemie International Edition. 2018;**57**(26):7790–7794. DOI: 10.1002/anie.201803222

[67] Shahbazi MA, Bauleth-Ramos T, Santos HA. DNA hydrogel assemblies: Bridging synthesis principles to biomedical applications. Advanced Therapeutics. 2018;**1**(4):1800042. DOI: 10.1002/adtp.201800042

[68] Wang Q, Qu YF, Zhang ZY, Huang H, Xu YF, Shen FY, et al. Injectable DNA hydrogel-based local drug delivery and immunotherapy. Gels. 2022;**8**(7):400. DOI: 10.3390/gels8070400

[69] Zhao J, Wu JY, Fan YQ, Yu C, Yu L, Shao FW. A pH-responsive dendritic-DNA-based nanohydrogel for dual drug delivery. Biomolecules. 2025;**15**(4):537. DOI: 10.3390/biom15040537

[70] Xing ZY, Caciagli A, Cao TY, Stoev I, Zupkauskas M, O'Neill T, et al. Microrheology of DNA hydrogels. Proceedings of the National Academy of Sciences of the United States of America. 2018;**115**(32):8137–8142. DOI: 10.1073/pnas.1722206115

[71] Wang YR, Zhu Y, Hu Y, Zeng GM, Zhang Y, Zhang C, et al. How to construct DNA hydrogels for environmental applications: Advanced water treatment and environmental analysis. Small. 2018;**14**(17):1703305. DOI: 10.1002/smll.201703305

[72] Arif M, Farooqi ZH, Irfan A, Begum R. Gold nanoparticles and polymer microgels: Last five years of their happy and successful marriage. Journal of Molecular Liquids. 2021;**336**:116270. DOI: 10.1016/j.molliq.2021.116270

[73] Cui JH, Wu D, Sun Q, Yang XZ, Wang DD, Zhuang M, et al. A PEGDA/DNA hybrid hydrogel for cell-free protein synthesis. Frontiers in Chemistry. 2020;**8**:28

[74] Chu J, Chen C, Li XJ, Yu LB, Li WJ, Cheng MX, et al. A responsive pure DNA hydrogel for label-free detection of lead ion. Analytica Chimica Acta. 2021;**1157**:338400. DOI: 10.1016/j.aca.2021.338400

[75] Huang YS, Xu WL, Liu GY, Tian LL. A pure DNA hydrogel with stable catalytic ability produced by one-step rolling circle amplification. Chemical Communications. 2017;**53**(21):3038–3041. DOI: 10.1039/c7cc00636e

[76] Hao LL, Wang W, Shen XQ, Wang SL, Li Q, An FL, et al. A fluorescent DNA hydrogel aptasensor based on the self-assembly of rolling circle amplification products for sensitive detection of ochratoxin A. Journal of Agricultural and Food Chemistry. 2020;**68**(1):369–375. DOI: 10.1021/acs.jafc.9b06021

[77] Li ZY, Davidson-Rozenfeld G, Vázquez-González M, Fadeev M, Zhang JJ, Tian H, et al. Multi-triggered supramolecular DNA/Bipyridinium dithienylethene hydrogels driven by light, redox, and chemical stimuli for shape-memory and self-healing applications. Journal of the American Chemical Society. 2018;**140**(50):17691–17701. DOI: 10.1021/jacs.8b10481

[78] Li J, Song WZ, Li F. Polymeric DNA hydrogels and their applications in drug delivery for cancer therapy. Gels. 2023;**9**(3):239. DOI: 10.3390/gels9030239

[79] Akintayo CO, Creusen G, Straub P, Walther A. Tunable and large-scale model network starPEG-DNA hydrogels. Macromolecules. 2021;**54**(15):7125–7133. DOI: 10.1021/acs.macromol.1c00600

[80] Chen ZM, Liu FY, Chen YK, Liu J, Wang XY, Chen AT, et al. Targeted delivery of CRISPR/Cas9-Mediated cancer gene therapy via liposome-templated hydrogel nanoparticles. Advanced Functional Materials. 2017;**27**(46):1703036. DOI: 10.1002/adfm.201703036

[81] Jin J, Kusamori K, Tanifuji T, Yamagata Y, Itakura S, Nishikawa M. Biocompatible DNA hydrogel composed of minimized Takumi-shaped DNA nanostructure exhibits sustained retention after *in vivo* administration. Journal of Controlled Release. 2025;**377**:704–715. DOI: 10.1016/j.jconrel.2024.11.052

[82] Jung IY, Kim JS, Choi BR, Lee K, Lee H. Hydrogel based biosensors for in vitro diagnostics of biochemicals, proteins, and genes. Advanced Healthcare Materials. 2017;**6**(12):1601475. DOI: 10.1002/adhm.201601475

[83] Zhang SB, Luan YX, Xiong MY, Zhang JJ, Lake R, Lu Y. DNAzyme amplified aptasensing platform for ochratoxin a detection using a personal glucose meter. ACS Applied Materials & Interfaces. 2021;**13**(8):9472–9481. DOI: 10.1021/acsami.0c20417

[84] Mann H, Khan S, Prasad A, Bayat F, Gu J, Jackson K, et al. Bacteriophage-Activated dnazyme hydrogels combined with machine learning enable point-of-use colorimetric detection of *Escherichia coli*. Advanced Materials. 2025;**37**(3):2411173. DOI: 10.1002/adma.202411173

[85] Chen MM, Wang Y, Zhao XD, Zhang JY, Peng Y, Bai JL, et al. Target-responsive DNA hydrogel with microfluidic chip smart readout for quantitative point-of-care testing of creatine kinase MB. Talanta. 2022;**243**:123338. DOI: 10.1016/j.talanta.2022.123338

[86] Khajouei S, Ravan H, Ebrahimi A. Developing a colorimetric nucleic acid-responsive DNA hydrogel using DNA proximity circuit and catalytic hairpin assembly. Analytica Chimica Acta. 2020;**1137**:1–10. DOI: 10.1016/j.aca.2020.08.059

[87] Liu C, Gou SY, Bi YH, Gao Q, Sun JJ, Hu SJ, et al. Smart DNA-gold nanoparticle hybrid hydrogel film based portable, cost-effective and storable biosensing system for the colorimetric detection of lead (II) and uranyl ions. Biosensors and Bioelectronics. 2022;**210**:114290. DOI: 10.1016/j.bios.2022.114290

[88] Umeki Y, Saito M, Kusamori K, Tsujimura M, Nishimura M, Takahashi Y, et al. Combined encapsulation of a tumor antigen and immune cells using a self-assembling immunostimulatory DNA hydrogel to enhance antigen-specific tumor immunity. Journal of Controlled Release. 2018;**288**:189–198

[89] Oda W, Umemura K, Ito K, Kawamoto Y, Takahashi Y, Takakura Y.

Development of potent unmethylated CpG DNA hydrogel by introducing i-motifs into long single-stranded DNA. International Journal of Pharmaceutics. 2023; **646**:123438. DOI: 10.1016/j.ijpharm.2023.123438

[90] Shao Y, Sun ZY, Wang YJ, Zhang BD, Liu DS, Li YM. Designable immune therapeutical vaccine system based on DNA supramolecular hydrogels. ACS Applied Materials & Interfaces. 2018;**10**(11):9310–9314. DOI: 10.1021/acsami.8b00312

[91] Fan Y, Zhan MS, Liang JH, Yang XS, Zhang BB, Shi XY, et al. Programming injectable DNA hydrogels yields tumor microenvironment-activatable and immune-instructive depots for augmented chemo-immunotherapy. Advanced Science. 2023;**10**(29):2302119. DOI: 10.1002/advs.202302119

[92] Wang FJ, Xie M, Huang YY, YH L, XL L, Zhu LJ, et al. In situ vaccination with an injectable nucleic acid hydrogel for synergistic cancer immunotherapy. Angewandte Chemie International Edition. 2024;**63**(4):e202315282. DOI: 10.1002/anie.202315282

[93] Wang ZH, Li W, Gou LP, Zhou Y, Peng G, Zhang JY, et al. Biodegradable and antioxidant DNA hydrogel as a Cytokine delivery system for diabetic wound healing. Advanced Healthcare Materials. 2022;**11**(21):2200782. DOI: 10.1002/adhm.202200782

[94] Qi C, Sun Q, Xiao DX, Zhang M, Gao SJY, Guo B, et al. Tetrahedral framework nucleic acids/hyaluronic acid-methacrylic anhydride hybrid hydrogel with antimicrobial and anti-inflammatory properties for infected wound healing. International Journal of Oral Science. 2024;**16**(1):30. DOI: 10.1038/s41368-024-00290-3

[95] Shen CY, Wang J, Li GF, Hao SY, Wu Y, Song PR, et al. Boosting cartilage repair with silk fibroin-DNA hydrogel-based cartilage organoid precursor. Bioactive Materials. 2024;**35**:429–444

[96] Wu X, Hu Y, Sheng SH, Yang HJ, Li ZH, Han QL, et al. DNA-based hydrogels for bone regeneration: A promising tool for bone organoids. Materials Today Bio. 2025;**31**:101502. DOI: 10.1016/j.mtbio.2025.101502

[97] Wei ZY, XX L, YC C, ZP H, Li Y, Gan L, et al. Programmable DNA-peptide conjugated hydrogel via click chemistry for sequential modulation of peripheral nerve regeneration. Advanced Functional Materials. 2025;**35**(23):2419915. DOI: 10.1002/adfm.202419915

[98] Wan L, Chen QS, Liu JB, Yang XH, Huang J, Li L, et al. Programmable self-assembly of DNA-Protein hybrid hydrogel for enzyme encapsulation with enhanced biological stability. Biomacromolecules. 2016;**17**(4):1543–1550. DOI: 10.1021/acs.biomac.6b00233

[99] Ma ZX, Wu Y, GF L, JL L, Geng Z, Su JC. Extracellular vesicles-loaded DNA hydrogels: A promising candidate for cartilage organoids engineering. Chemical Engineering Journal. 2023; **477**:147146. DOI: 10.1016/j.cej.2023.147146

[100] Yang B, Zhou BN, Cf L, XW L, Shi ZW, Li YX, et al. A biostable l-DNA hydrogel with improved stability for biomedical applications. Angewandte Chemie International Edition. 2022;**61**(30):e202202520. DOI: 10.1002/anie.202202520

[101] Kim N, Lee H, Han G, Kang M, Park S, Kim DE, et al. 3D-Printed functional hydrogel by DNA-Induced

biomineralization for accelerated diabetic wound healing. Advanced Science. 2023;**10**(17):2300816. DOI: 10.1002/advs.202300816

[102] Guo WW, Lu CH, Orbach R, Wang FA, Qi XJ, Cecconello A, et al. pH-Stimulated DNA hydrogels exhibiting shape-memory properties. Advanced Materials. 2015;**27**(1):73–78. DOI: 10.1002/adma.201403702

[103] Nayak S, Prasad SR, Mandal D, Das P. Hybrid DNA-carbon dot-poly(vinylpyrrolidone) hydrogel with self-healing and shape memory properties for simultaneous trackable drug delivery and visible-light-induced antimicrobial photodynamic inactivation. ACS Applied Bio Materials. 2020;**3**(11):7865–7875. DOI: 10.1021/acsabm.0c01022

[104] Han JP, Cui YC, Han XP, Liang CY, Liu WG, Luo D, et al. Super-soft DNA/dopamine-grafted-dextran hydrogel as dynamic wire for electric circuits switched by a microbial metabolism process. Advanced Science. 2020;7(13):2000684. DOI: 10.1002/advs.202000684

[105] Bi YH, Du XX HPP, Wang CY, Liu C, Guo WW. Smart bilayer polyacrylamide/DNA hybrid hydrogel film actuators exhibiting programmable responsive and reversible macroscopic shape deformations. Small. 2020;**16**(42):1906998. DOI: 10.1002/smll.201906998

[106] Dave N, Chan MY, Huang PJJ, Smith BD, Liu JW. Regenerable DNA-Functionalized hydrogels for ultrasensitive, instrument-free mercury(II) detection and removal in water. Journal of the American Chemical Society. 2010;**132**(36):12668–12673. DOI: 10.1021/ja106098j

[107] Zhu XL, Mao XX WZH, Feng C, Chen GF, Li GX. Fabrication of nanozyme@DNA hydrogel and its application in biomedical analysis. Nano Research. 2017;**10**(3):959–970. DOI: 10.1007/s12274-016-1354-9

[108] Jiang HT, Qi S, Dong XZ, Ang B, Zhang Y, Wang ZP. Integrating metal-organic framework hybrid into nucleic acid-based hydrogel for highly selective recognition and sensitive detection of sarafloxacin. Analytical Chemistry. 2025;**97**(20):10550–10560. DOI: 10.1021/acs.analchem.4c05180

[109] Zhang LB, Jean SR, Ahmed S, Aldridge PM, Li XY, Fan FJ, et al. Multifunctional quantum dot DNA hydrogels. Nature Communications. 2017;**8**:381

[110] Lemke P, Schneider L, Kunz W, Rieck AL, Jäger PS, Bruckmann A, et al. Flow-Induced microfluidic assembly for advanced biocatalysis materials. Advanced Functional Materials. 2024;**34**(18):2313944. DOI: 10.1002/adfm.202313944

[111] Gao M, Krissanaprasit A, Miles A, Hsiao LLC, LaBean TH. Mechanical and electrical properties of DNA hydrogel-based composites containing self-assembled three-dimensional nanocircuits. Applied Sciences. 2021;**11**(5):2245. DOI: 10.3390/app11052245

[112] Hu YW, Ying JY. Reconfigurable A-motif, i-motif and triplex nucleic acids for smart pH-responsive DNA hydrogels. Materials Today. 2023;**63**:188–209. DOI: 10.1016/j.mattod.2022.12.003

[113] Lu SS, Wang S, Zhao JH, Sun J, Yang XR. A pH-controlled bidirectionally pure DNA hydrogel: Reversible self-assembly and

fluorescence monitoring. Chemical Communications. 2018;**54**(36): 4621–4624. DOI: 10.1039/c8cc01603h

[114] Chen ZY, Zhang H, Huang JT, Weng WZ, Geng Z, Li MM, et al. DNA-encoded dynamic hydrogels for 3D bioprinted cartilage organoids. Materials Today Bio. 2025;**31**:101509. DOI: 10.1016/j.mtbio.2025.101509

[115] Morikawa K, Masubuchi Y, Shchipunov Y, Zinchenko A. DNA-Chitosan hydrogels: Formation, properties, and functionalization with catalytic nanoparticles. ACS Applied Bio Materials. 2021;**4**(2):1823–1832. DOI: 10.1021/acsabm.0c01533

[116] Zinchenko A, Miwa Y, Lopatina LI, Sergeyev VG, Murata S. DNA hydrogel as a template for synthesis of ultrasmall gold nanoparticles for catalytic applications. ACS Applied Materials & Interfaces. 2014;**6**(5):3226–3232. DOI: 10.1021/am5008886

[117] Fadeev M, Davidson-Rozenfeld G, Li ZZ, Willner I. Stimuli-responsive DNA-based hydrogels on surfaces for switchable bioelectrocatalysis and controlled release of loads. ACS Applied Materials & Interfaces. 2023;**15**(30):37011–37025. DOI: 10.1021/acsami.3c06230

[118] Zinchenko A, Che YX, Taniguchi S, Lopatina LI, Sergeyev VG, Murata S. Metallization of DNA hydrogel: Application of soft matter host for preparation and nesting of catalytic nanoparticles. Journal of Nanoparticle Research. 2016;**18**(7):179

[119] Gaspers P, Bickmann C, Wallner C, Diaz DB, Holtmann D, Gescher J, et al. Extracellular bacterial production of DNA hydrogels-toward engineered living materials. Small. 2025;**21**(19):2502199. DOI: 10.1002/smll.202502199

[120] Tang JP, Wang J, Ou JH, Cui Z, Yao C, Yang DY. A DNA/Poly-(L-lysine) hydrogel with long shelf-time for 3D cell culture. Small Methods. 2024;**8**(7):2301236. DOI: 10.1002/smtd.202301236

Chapter 4

Smart Hydrogel-Based Actuators and Artificial Muscles: Emerging Frontiers in Soft Robotics

Nazia Mustary Nayna and Sabrina Sabnam Samonty

Abstract

Smart hydrogel-based actuators have emerged as a prominent class of soft robotic components due to their intrinsic compliance, biocompatibility, and multi-stimuli responsiveness. These materials can undergo reversible deformations—including bending, twisting, swelling, contraction, and locomotion—when exposed to thermal, electrical, chemical, magnetic, or optical stimuli, enabling close emulation of natural muscle behavior. Owing to their high water content and tunable polymer networks, hydrogel actuators offer safe and adaptive interfaces for biological and human-centric applications. Recent advances in polymer chemistry, nanocomposite reinforcement, additive manufacturing, and biohybrid design have significantly expanded the performance envelope of hydrogel actuators. Innovations in fabrication strategies such as 3D/4D printing, micro-patterning, and layered architectures have enabled precise control over actuation modes, response speed, and force generation. As a result, hydrogel actuators are increasingly explored in artificial muscles, biomedical microrobots, implantable devices, wearable systems, and intelligent soft robotic platforms. This chapter presents a comprehensive review of hydrogel fundamentals, actuation mechanisms, material systems, and fabrication approaches, followed by an overview of key application domains. Comparative benchmarking with alternative artificial muscle technologies is provided, highlighting performance trade-offs in strain, stress output, efficiency, and controllability. Emerging integration of sensing and self-feedback functionalities is also discussed. Finally, critical challenges—including dehydration, limited mechanical strength, energy efficiency, and scalability—are examined, with future perspectives emphasizing biohybrid systems, nanocomposite hydrogels, and 4D-printed architectures for next-generation intelligent soft robots. If you want, I can adapt this for a PhD thesis chapter, Springer/Elsevier book chapter, or IEEE-style review with references added.

Keywords: hydrogels, soft robotics, artificial muscles, stimuli-responsive polymers, additive manufacturing, biohybrid actuators, 4D printing, nanocomposite hydrogels, wearable robotics, biomedical microrobots

1. Introduction

Soft robotics has emerged as a transformative field, bridging biology, materials science, and robotics, with the aim of developing machines that are safer, more adaptable, and capable of mimicking natural organisms. Among the various materials investigated, hydrogels have gained particular importance due to their high-water content, tissue-like compliance, tunable mechanical properties, and ability to respond to multiple external stimuli. These characteristics make hydrogels one of the most promising classes of smart materials for actuators and artificial muscles, particularly in biomedical and bioinspired applications where softness, flexibility, and biocompatibility are critical.

The academic value of this chapter lies in providing a comprehensive and structured overview of **hydrogel-based actuators as emerging frontiers in soft robotics**. While numerous reviews exist on stimuli-responsive hydrogels, relatively few provide an integrated perspective that connects fundamental material design, actuation mechanisms, fabrication strategies, sensing integration, and real-world applications in artificial muscles. This chapter addresses this gap by benchmarking hydrogels against other artificial muscle technologies and highlighting the progress toward multifunctional, intelligent, and feedback-controlled hydrogel systems.

The content is structured as follows:

(i) Fundamentals of hydrogel actuation and the chemistry underlying their responsiveness;

(ii) Classification of actuation mechanisms and associated deformation modes;

(iii) Applications of hydrogels as artificial muscles and their comparison with other actuator technologies;

(iv) Integration of sensing with actuation for intelligent systems;

(v) Advances in fabrication strategies, including additive manufacturing and bio fabrication;

(vi) Applications in biomedical robotics, microrobots, and wearable systems;

(vii) Emerging frontiers in biohybrid systems, nanocomposites, and 4D printing; and

(viii) Challenges and future outlook for transitioning hydrogel actuators into scalable, real-world technologies.

This framework not only situates hydrogels within the broader field of soft robotics but also provides insights into their transformative potential in next-generation robotic and biomedical systems. Recent studies have emphasized the growing role of smart hydrogels in artificial muscles, with developments in double-network structures, nanocomposite hydrogels, and 4D-printed architectures enabling improved stress output, faster response times, and enhanced durability [1–3].

2. Fundamentals of hydrogel actuation in soft robotics

Hydrogels are networks of hydrophilic polymers capable of retaining large amounts of water, giving them a tissue-like softness and elasticity. These materials closely mimic the mechanical characteristics of natural muscle and other soft tissues, making them highly bioinspired candidates for soft robotics applications [1]. Unlike traditional silicone elastomers (e.g., PDMS) commonly used in soft robots, hydrogels offer **inherent biocompatibility, tunable responsiveness to stimuli, and the ability to serve as both structural and functional elements** (actuators or sensors) simultaneously [1, 4]. For instance, a hydrogel can swell or shrink in response to environmental changes, directly converting chemical or physical stimuli into mechanical motion, an example of *embodied intelligence* not easily achieved with conventional rigid actuators.

Hydrogel actuators operate primarily via volumetric changes (swelling/de-swelling) induced by stimuli. The high water content (often > 90%) enables fast shape changes by water absorption or expulsion within the polymer network [4]. This mechanism is analogous to how natural muscles change shape by modulating internal fluid content. Early research on hydrogels focused on biomedical uses (drug delivery, tissue engineering), but as their unique properties became evident, they gained attention in soft robotics as **"smart" materials** that can respond autonomously to stimuli like temperature, pH, electric fields, light, and magnetic fields [3]. The appeal of hydrogels lies in features such as **soft compliance**, **high deformability**, and **intrinsic sensing** (electrical or optical property changes upon deformation), which are crucial for safe interaction with humans and adaptive robot behaviors.

Notably, the past few decades have seen accelerating interest in hydrogel-based soft actuators, referencing exponential growth in publications related to stimuli-responsive hydrogels since the 1990s, reflecting their emerging importance in soft robotics research [4]. This trend parallels the rise of soft robotics as a field, indicating that researchers are increasingly exploring hydrogels as alternatives to or integrative components with elastomers. However, despite these advantages, hydrogels historically represent only a small fraction of soft robotics materials due to several challenges (discussed later), such as mechanical fragility and the need to remain hydrated [1]. Overcoming these hurdles is a key focus of current research to fully unlock hydrogels' potential in creating soft actuators and artificial muscles.

Figure 1 illustrates the exponential growth in publications related to stimuli-responsive hydrogels since the 1990s, highlighting their increasing importance in soft robotics research.

3. Actuation mechanisms of hydrogel-based soft actuators

Hydrogel actuators can be categorized by both **stimulus modality** and the resulting **motion type** they produce. Virtually any type of actuation stimulus can, in principle, yield any basic mode of motion (expansion, contraction, bending, twisting, folding), though certain stimulus–motion combinations are more practical than others [1]. The actuation mechanism in each case relies on a physicochemical response of the hydrogel network to an external cue, converting stimuli into mechanical work through swelling, shrinking, or phase change. Key actuation strategies include:

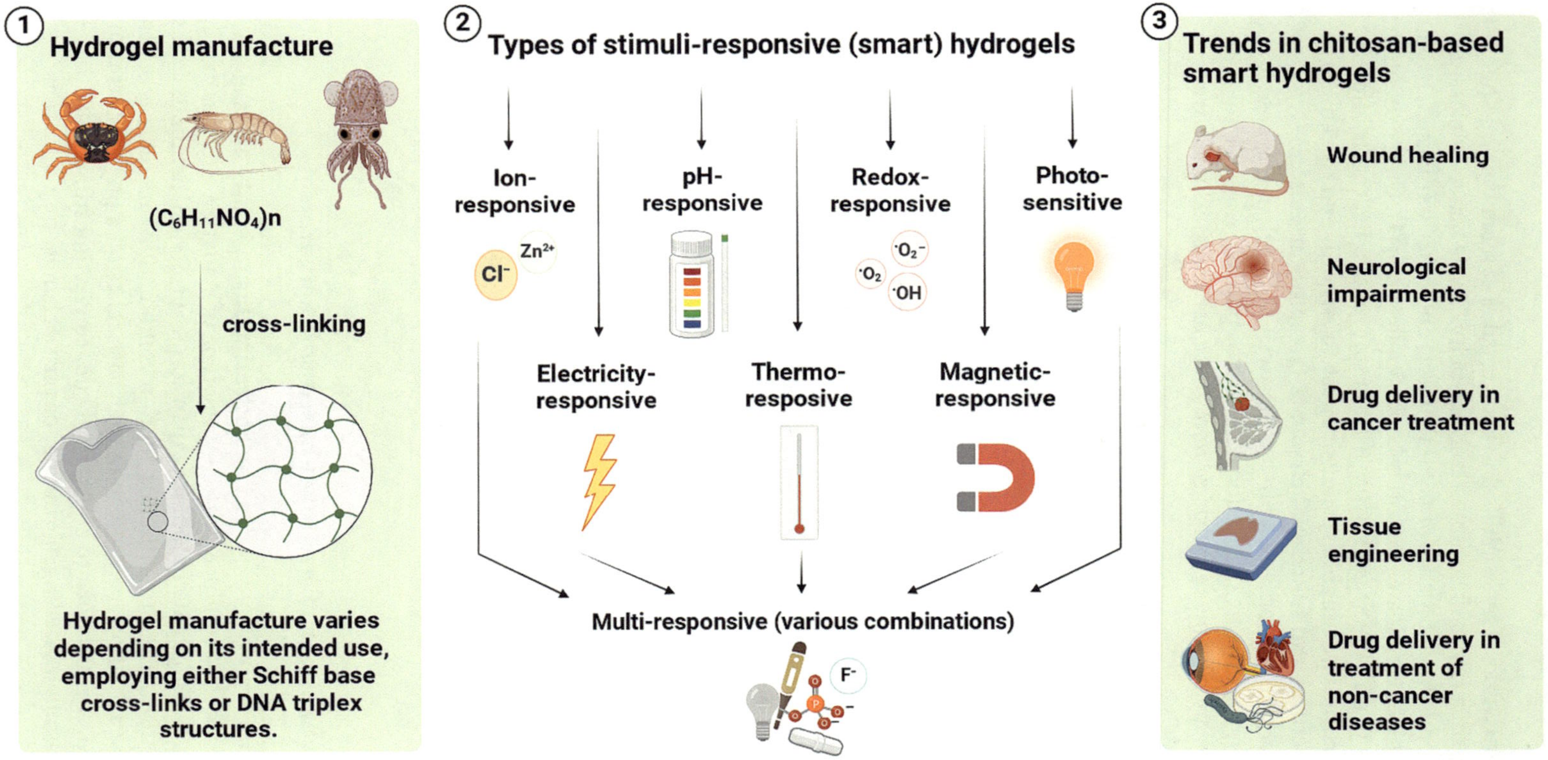

Figure 1.
An overview of hydrogel manufacture, stimuli responsiveness, and emerging biomedical applications.
Manufacture, stimuli-responsiveness, and biomedical trends of smart hydrogels.
(1) Hydrogel manufacture: derived from natural sources (e.g., Chitosan, squid cartilage) via cross-linking strategies such as Schiff base and DNA triplex structures.
(2) Types of smart hydrogels: ion-responsive, pH-responsive, redox-responsive, photosensitive, electricity-responsive, thermo-responsive, magnetic-responsive, and multi-responsive combinations.
(3) Emerging biomedical trends in chitosan-based hydrogels include wound healing, neurological impairment therapies, drug delivery [cancer and non-cancer], and tissue engineering. Adapted from [Frontiers in Chemistry, [5, 6] 2021; 1] in summary, hydrogel-based actuators offer a ***bioinspired approach to soft robot design*** *by directly harnessing material intelligence. Their ability to deform massively and reversibly under mild stimuli, much like living tissues, makes them prime candidates for next-generation* soft artificial muscles. *In the following sections, we discuss the various actuation mechanisms employed by hydrogel actuators, their integration as artificial muscles, advances in sensing and fabrication, and the emerging frontiers and challenges in this exciting area of soft robotics.*

- **Thermal actuation:** Hydrogels formulated with a Lower Critical Solution Temperature (LCST) (e.g., PNIPAM) undergo pronounced volume changes when crossing a threshold temperature. Heating such a hydrogel above its LCST causes it to expel water and collapse (shrinking), whereas cooling allows it to reswell. This volume phase transition can produce expansive or contractive motion. Thermally-responsive hydrogel actuators tend to operate with a *smooth but slow response* and are often used in devices like temperature-controlled valves, artificial muscles, and drug delivery systems [4]. A classic example is a bilayer of PNIPAM that bends upon heating, mimicking plant leaf movements. Thermal actuation is attractive for its simplicity but can be sluggish and may require significant temperature changes (which, in biomedical contexts, must remain within safe limits).

- **Electric field (electroactive) actuation:** Hydrogels containing ionic groups respond to electric fields by redistributing ions and creating osmotic swelling gradients. In an **ionic electroactive hydrogel** (a type of ionic Polymer–Metal composite when electrodes are attached), an applied voltage drives cations and anions within the gel, causing one side to swell more than the other – resulting in bending toward one electrode [1]. Early work by Shiga (1990) showed that polyacrylate hydrogels bend in salt solution under a DC voltage, pioneering the concept of **artificial hydrogel muscles powered by electricity**. These electrically-driven hydrogel actuators offer relatively fast response and can be precisely controlled by voltage, but historically, they needed to remain hydrated in an electrolyte solution. Recent advances include hydrogels that retain water internally or are coated to prevent drying, enabling operation in air [1]. Electroactive hydrogels can achieve fine movements (crawling, inching motions, etc.) and are being explored for electrically controlled soft robotic manipulators. They respond faster than many other stimuli (with response times on the order of seconds), and the development of conductive nanocomposites (e.g., graphene-infused hydrogels) has further improved their speed and deformation range [1].

- **Magnetic field actuation:** By embedding magnetic nanoparticles or ferromagnetic microparticles into a hydrogel, the composite can be actuated remotely by an external magnetic field. The field exerts forces/torques on the particles, yielding bulk deformation of the soft matrix. Magnetic-responsive hydrogels have the advantage of *wireless control* and rapid response. For example, paramagnetic particle-loaded hydrogels can shrink or bend within seconds of field application, and can even exert significant forces (useful for lifting or moving objects in a robotic context) [4]. One study demonstrated a hydrogel with Fe nanoparticle fillers that achieved fast, reversible shape changes under an oscillating magnetic field, behaving like an artificial muscle with contractions and relaxations. A challenge for magnetic hydrogel actuators is that adding sufficient magnetic filler can make the hydrogel mechanically stiffer or more brittle, and such composites can be difficult to fabricate via printing methods (due to particle sedimentation or nozzle clogging) [4]. Nonetheless, magnetic actuation is highly promising for untethered soft robots (for instance, millimeter-scale hydrogel swimmers or crawlers driven by magnetic coils).

- **Light actuation (photo-responsive):** Hydrogels can be formulated or doped with photochromic molecules, photothermal nanoparticles, or other light-sensitive components so that illumination triggers a volume change. For instance, hydrogels containing azobenzene or spiropyran undergo conformational changes upon UV or visible light exposure, leading to either hydrophilicity changes (and thus swelling) or direct mechanical deformation. An example is a PNIPAM hydrogel with gold nanorods: under near-infrared light, the nanorods heat the hydrogel locally (photothermal effect), causing it to shrink and bend; when the light is off, the hydrogel cools and re-swells, allowing cyclic motion. Light offers high spatial and temporal control (even allowing remote, non-contact actuation), but a limitation is the penetration depth – hydrogels thicker than a few millimeters may not receive uniform light stimulation throughout their volume, and some photosensitive chemistries require UV light, which is less practical for in vivo or safe robotics applications [4]. Despite these challenges, light-driven hydrogel actuators have been demonstrated in applications like micro-valves, soft micro-robots, and tunable lenses, and they enable truly untethered operation (no wires or tubes).

- **Chemical or solvent-driven actuation:** Hydrogels also respond to chemical stimuli such as pH, specific ions, or solvent composition. For example, a polyacid hydrogel will dramatically swell when the pH rises above its pKa (due to ionization of pendant groups causing electrostatic repulsion) – this principle can drive pH-sensitive gripping or releasing actions. Similarly, hydrogels that are responsive to glucose or other biomolecules have been explored for biomimetic actuators (though more commonly for sensing). Another important class is **humidity-driven actuators**, where a desiccated hydrogel film will expand by absorbing moisture. Researchers have created humidity-responsive hydrogel films that curl or oscillate with changes in ambient humidity, useful for soft hygrometers or self-wrapping devices. These chemically-triggered actuation modes tend to be slower, as they rely on the diffusion of molecules into/out of the gel, but they can be highly useful for *environmental robotic sensors* or autonomous environmental controls (e.g., a hydrogel that opens a vent when humidity rises) [3].

- **Pneumatic and hydraulic actuation:** In some cases, hydrogels are used in more traditional fluidic actuators – for instance, as the elastomeric bladder in a soft pneumatic muscle. One novel concept involves using a **hydrogel as a soft hydraulic fluid**: Recent work by Kim et al. created a "hydraulic artificial muscle" where an electrically heated dielectric elastomer pump drives fluid pressure inside a tubular hydrogel, causing it to expand and contract like a McKibben muscle [3]. In other designs, dehydration of a hydrogel under negative pressure can mimic the volume change of a pneumatic actuator. The advantage lies in combining the large strain of fluidic actuation with the damping and compliance of hydrogels. However, these hybrids still require external pumps or pressure sources, so the trend is toward stimuli-responsive hydrogels that eliminate the need for rigid pneumatic hardware.

Table 1 summarizes some of the major stimuli used in hydrogel actuators, highlighting their key characteristics and typical application areas [4]. Each

Stimulus	Mechanism	Response speed	Pros	Cons	Uses
Thermal (e.g., PNIPAm LCST)	Volume phase transition	Slow–moderate	Large strain, simple trigger	Heating overhead, slow	Self-folding, valves, drug release [4]
Electrical (ionic/ electroactive)	Ion migration → osmotic gradient	Fast	Precise control	Needs hydration	Artificial muscles, bending actuators [1]
Magnetic (NP-filled)	Torque/force on fillers	Fast	Wireless control	Particle stiffening	Untethered crawlers, targeted devices [4]
Light (photo/ photothermal)	Isomerization/ local heating	Fast (local)	Non-contact, selective	Limited penetration	Microvalves, microrobots [4]
Chemical/pH/ solvent	Ionization/ swelling	Slow	Environment-coupled	Diffusion-limited	Environmental robots [3]
Pneumatic/ Hydraulic	Pressure-driven deformation	Fast strong	High force/ strain	Needs pumps	PAM-like actuators [7]

Table 1.
Common stimuli for hydrogel-based actuation, with their characteristics and representative application areas. Hydrogels translate various energy inputs into mechanical motion, enabling soft robots that are triggered by temperature, electric fields, magnetic fields, or light [4].

actuation mode has distinct advantages and constraints, and researchers are actively developing multimodal hydrogels that respond to multiple stimuli (e.g., a hydrogel that is both photo- and thermo-responsive) to create more versatile soft actuators.

The key material, actuation, and application-level challenges associated with smart hydrogel-based actuators are summarized in **Figure 2**. As shown in **Figure 2**, the challenges of smart hydrogel-based actuators can be grouped into material and manufacturing limitations, actuation and integration constraints, and application-level issues. Beyond single-mode actuators, researchers are now exploring **multimodal actuation** in hydrogels – for example, gels that respond to both pH and temperature, or systems combining an internal chemical reaction with external magnetic guidance. Such actuators could switch modes or even have one stimulus trigger a gross motion, and another fine-tune the position. An emerging concept is to integrate multiple hydrogel actuators, each tuned to different stimuli, in one device (so one robot can be controlled by, say, temperature *and* light in different scenarios) [6]. This multimodal approach can increase the versatility and adaptiveness of soft robots, at the cost of added material complexity.

In summary, hydrogel-based actuators encompass a rich variety of mechanisms. **Their motions are fundamentally driven by swelling mechanics**, but through clever chemistry and material design, this simple principle translates into diverse actuation behaviors. The proper choice of stimulus and hydrogel chemistry allows engineers to tailor actuator performance: *fast electrical twitches, strong magnetic pulls, gentle thermally-induced shape changes, or precise light-guided motions* are all achievable with hydrogels. In the next section, we discuss how these actuation capabilities position hydrogels as promising **artificial muscles** in soft robotics, and how they compare or integrate with other artificial muscle technologies.

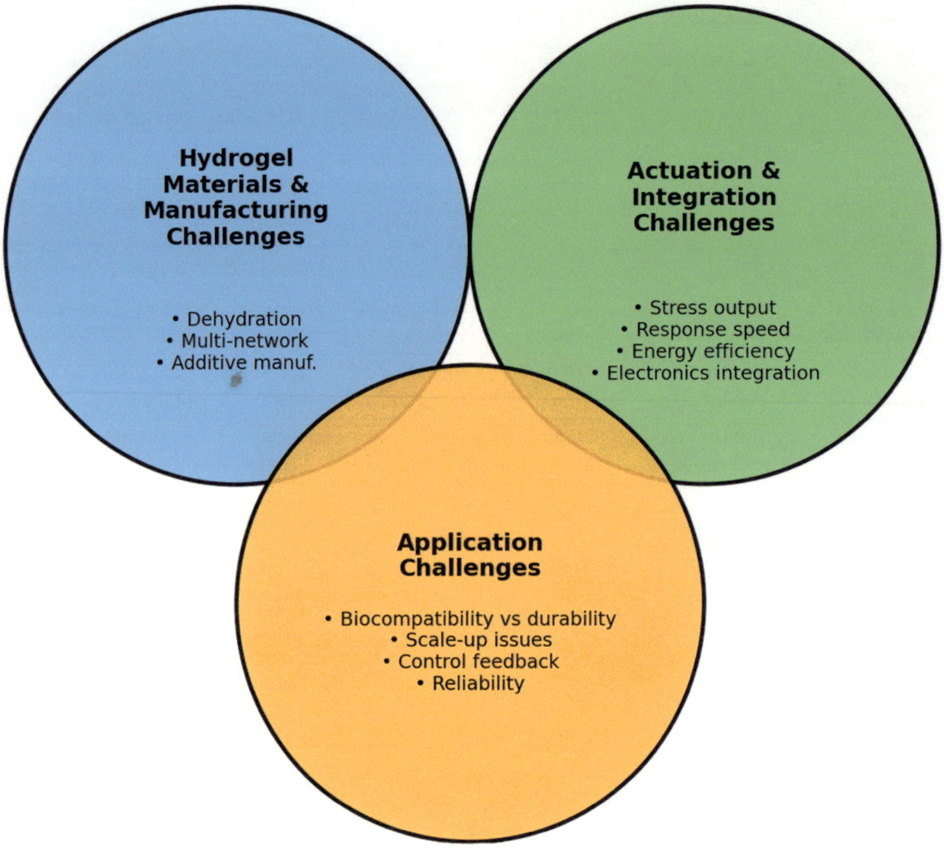

Figure 2.
Challenges in smart hydrogel-based actuators.
The challenges are grouped into three domains: (i) hydrogel materials and manufacturing (dehydration, multi-network design, additive manufacturing), (ii) actuation and integration (stress output, response speed, energy efficiency, electronics integration), and (iii) application-level issues (biocompatibility vs. Durability, scalability, control and feedback, and long-term reliability).

4. Hydrogel actuators as artificial muscles

"Artificial muscles" refer to materials or devices that reproduce the contractile behavior of natural muscle – namely, the ability to **reversibly contract, produce force, and do mechanical work** in response to a stimulus. Traditional artificial muscles include pneumatic McKibben actuators, shape memory alloys, dielectric elastomers, and polymer fibers (e.g., nylon fishing line actuators). Hydrogel-based actuators are a relatively newer addition to this roster, bringing unique benefits such as softness and biocompatibility, albeit at the expense of speed or strength in some cases. Recent bibliometric analyses show that research into artificial muscles is rapidly expanding, with novel materials like carbon nanotubes, graphene, and hydrogels being actively studied to improve performance [2]. In particular, hydrogels have attracted interest for bio-integrated artificial muscles (e.g., implantable assistive devices or prosthetics) due to their tissue-like characteristics [1].

One approach to create muscle-like actuation from hydrogels is to design **fiber or filament-shaped hydrogel actuators**. For example, Hu et al. (2021) [8], developed a **twisted and coiled hydrogel fiber** that functions analogously to a skeletal muscle fiber [3]. This hydrogel fiber was prepared with an internal electrolyte and wrapped with a helical constraint; when a voltage is applied, electro-osmotic flow causes the fiber to shrink in length (much like a muscle contraction) and then relax when the voltage is removed. Such fibers can be bundled in parallel to amplify force, mimicking muscle fascicles. The advantage is extremely high compliance (the fibers are as soft as the surrounding tissue) and inherent damping, although their contracting speed and generated stress currently lag behind human muscle. Nonetheless, these hydrogel fibers demonstrate that **linear contraction artificial muscles** can be built from hydrogels, not just bending actuators. An example of light-driven hydrogel actuation exhibiting jumping and rolling locomotion is shown in **Figure 3**.

Hydrogel actuators can also be arranged in **antagonistic pairs or networks** to achieve more complex motions, like muscle groups. For instance, researchers have combined an anionic (negative) hydrogel and a cationic (positive) hydrogel in a bilayer or parallel configuration; under an electric field, one gel swells while the other shrinks, creating a robust bending or linear motion. López-Díaz et al. [1] describe a walker robot composed of such ionic hydrogel pairs that can "walk" underwater when an oscillating electric field is applied – a simplistic but biomimetic locomotion strategy. Although the motion was slow (on the order of millimeters per minute), it highlights how **asymmetric swelling of multiple hydrogel elements** can be orchestrated for gait generation, analogous to how muscles coordinate for locomotion [1].

Table 2 Comparative properties of artificial muscle technologies. Hydrogels demonstrate outstanding compliance and biocompatibility but lag in stress output and fatigue resistance compared with dielectric elastomers, shape memory alloys, and pneumatic artificial muscles. Adapted from [1, 4, 10–15].

In comparison to established artificial muscle technologies, hydrogels currently face some performance limitations. Natural skeletal muscle can contract ~20% of its length in a fraction of a second and exert considerable force (~0.3 MPa stress). Hydrogel actuators typically have large strain potential (often swelling to several times their dry length, far exceeding muscle strain), but their **stress output is modest, and response times are slower** (seconds to minutes for full swelling, depending on size) [1]. However, research is actively addressing these gaps: Adding nanofillers (graphene, carbon nanotubes) has been shown to improve hydrogel mechanical strength and response speed (e.g., graphene-enhanced hydrogels respond faster to electric stimuli by improving electrical conductivity) [1]. Double-network hydrogels and other tough hydrogels have achieved greatly improved tensile strength and toughness, approaching that of load-bearing tissues [4]. By integrating such tough hydrogels into actuators, the force output increases. For example, magnetic hydrogel actuators containing iron microparticles have generated stresses on the order of tens of kPa and lifted objects several times their own weight under a magnetic field – performance that begins to approach pneumatic artificial muscles, but with much more compliance.

Another strategy is the **hierarchical structuring** of hydrogel actuators to mimic muscle architecture. Instead of a single bulk hydrogel, researchers create micro-structured hydrogel composites: fibers, sheets, or lamina that buckle, fold, or pull in unison. A bioinspired design by Kim et al. involved a *fibrous hydrogel composite* with anisotropic swelling – upon stimulation, the fibers contract along one axis but not

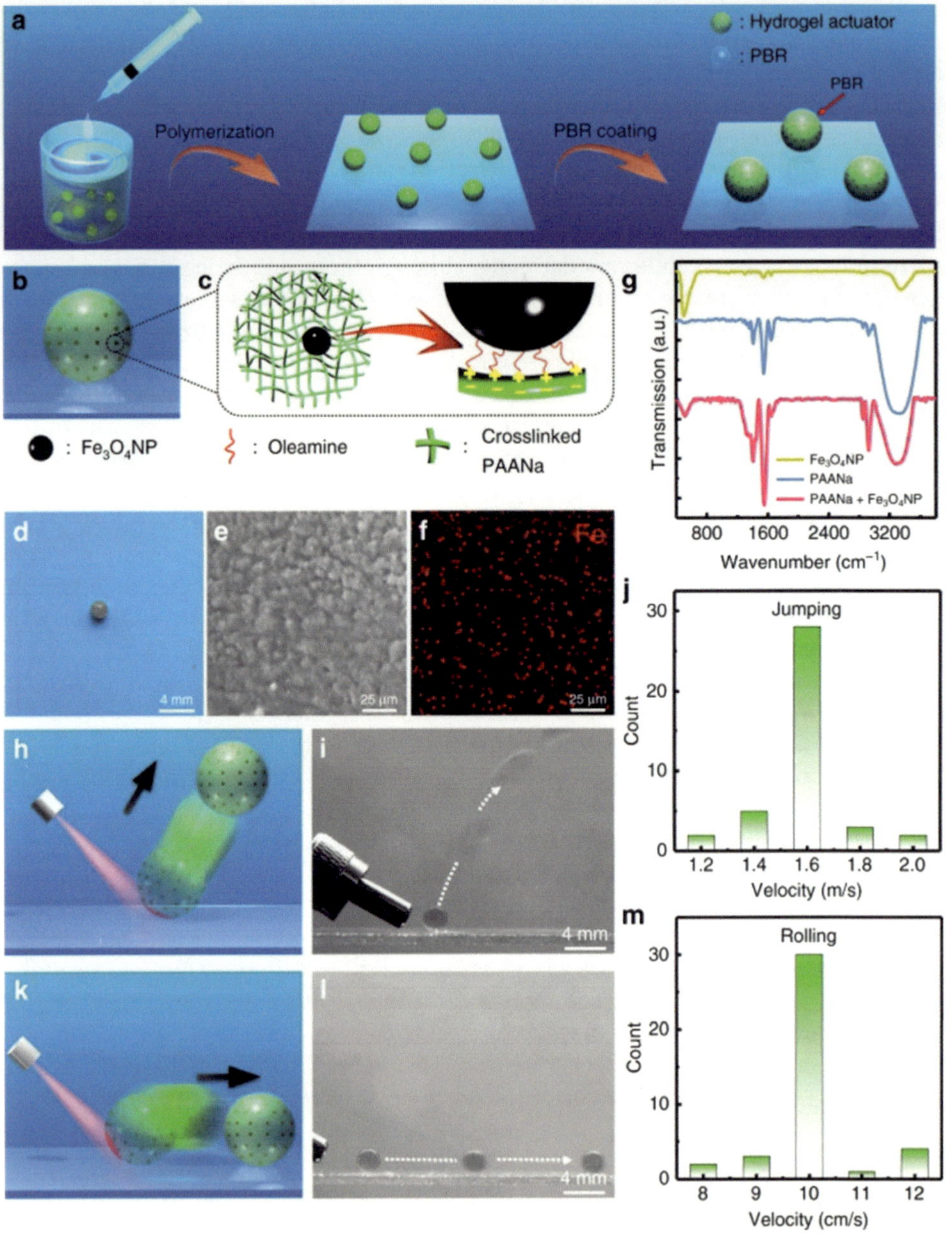

Figure 3.

The hydrogel actuator and its light-induced jumping and rolling motion in air. The device, composed of poly (sodium acrylate) hydrogel embedded with iron oxide nanoparticles, demonstrates ultrafast actuation – attaining speeds up to 1.6 m/s, response times as low as 800 ms, and jumping heights around 15 cm – powered by a photothermal bubble mechanism under NIR light irradiation [9].

the other, yielding a unidirectional contraction similar to muscle fiber bundles [4]. Such designs significantly increase the efficiency and directionality of hydrogel actuation.

Despite not yet matching the raw speed of some other artificial muscles, **hydrogels excel in their gentle touch and biocompatibility**. This makes them

Actuator type	Max strain (%)	Stress output (MPa)	Response speed	Biocompatibility	Fatigue resistance	Notes
Hydrogel actuators	30–200	0.01–0.1	Slow–Moderate	Excellent	Low–Moderate	Highly biocompatible; issues with dehydration; tunable via multiple stimuli [1,4].
Dielectric elastomer actuators (DEA)	10–300	1–8	Fast (ms–s)	Moderate	High	Large strain and fast actuation require high voltages (kV range) [9, 11].
Shape memory alloys (SMA)	3–8	200–500	Slow (s)	Low–Moderate	Moderate	Very high stress output; limited strain; heating/cooling required [11, 13].
Pneumatic artificial muscles (PAM)	20–30	0.3–0.8	Fast (ms–s)	Low–Moderate	High	High power-to-weight ratio; requires bulky compressors [14, 15].

Table 2.
Comparative performance of artificial muscle technologies.

extremely appealing for applications like medical robotics and assistive devices. One can envision hydrogel artificial muscle patches directly interfacing with organs (e.g., a cardiac assist device), where their softness and wetness are critical, or as actuators in wearable exosuits, where hard actuators would cause discomfort. Indeed, some recent studies have demonstrated **hydrogel-based actuating wound dressings and compresses** that dynamically change shape to apply therapeutic pressure or motion to an injury site – functions that would be difficult to achieve with traditional actuators.

In summary, hydrogel actuators are emerging as a compelling form of artificial muscle for scenarios where *safety, compliance, and biocompatibility* are paramount, and where moderate speeds and forces are acceptable. As materials science advances to create faster, stronger hydrogels (e.g., via nanocomposites or fiber reinforcement) [2], we can expect hydrogel artificial muscles to improve to the point of competing with or complementing existing artificial muscle technologies. The true power of hydrogels may ultimately lie in **hybrid systems** – for example, a hydrogel artificial muscle embedded with rigid fiber reinforcements or combined with an elastomer skin, achieving a balance of strength and softness. The next section delves into how hydrogels can also serve a dual role as sensors, paving the way for integrated actuation-sensing muscle systems.

5. Integration of sensing with hydrogel actuators

Soft robots will require not only muscle-like actuation but also nervous system-like sensing to achieve intelligent behavior. Hydrogels offer opportunities for

integrated sensing because many hydrogels are inherently responsive to stimuli in ways that can be transduced into electrical or optical signals. For instance, a hydrogel's electrical resistance, capacitance, or fluorescence can change with deformation, temperature, or chemical environment. This means a hydrogel actuator can often serve as its own sensor – a concept sometimes called "self-sensing" or proprioceptive actuation.

Researchers are actively exploring hydrogels as stretchable sensors for strain, pressure, humidity, and more. *Hydrogel-based strain sensors* can be made by forming a conductive hydrogel (e.g., a polyacrylamide network infused with salt or conductive nanoparticles); as the hydrogel stretches, its resistance increases, providing a readout of deformation. Wang et al. [3] review numerous examples of such sensors, noting that hydrogels have been used to detect human motion (as wearable sensors in gloves, e-skin patches, etc.) due to their skin-like softness and high stretchability. For example, an ionic hydrogel strain sensor attached to a finger can detect bending motions by the change in ionic current, with the hydrogel remaining transparent and comfortable on the skin [3]. Because hydrogels can conform intimately to surfaces, they achieve high sensitivity and a broad sensing range (strains up to several hundred percent in some cases).

Integrating sensing into hydrogel actuators can be done in a few ways:

- **Material-intrinsic sensing:** The hydrogel actuator itself is made conductive or otherwise sensing-capable. For example, a **conductive hydrogel muscle** can act as an actuator (contracting under stimuli), and simultaneously, its resistance provides feedback on the degree of contraction or strain. Lyu et al. have demonstrated a hydrogel that changes electrical resistance with expansion; by measuring this change, they closed the loop to control the actuator's position in real time [Zhou et al., 2024].

- **Embedded sensor networks:** Tiny sensors (like micro-LEDs, fiber optics, or microcapsules) can be embedded throughout a hydrogel. One example is a hydrogel with embedded Bragg grating optical fibers that reflect light according to strain – effectively creating a muscular optical sensor. Another example is encapsulating microfluidic channels of liquid metal inside a hydrogel actuator; stretching the hydrogel changes the geometry of the liquid metal circuit, altering its inductance or resistance, and thus sensing deformation. These approaches maintain the actuator's softness while adding sensing functionality.

- **Surface sensor integration:** Attaching thin-film sensors onto the hydrogel's surface. Although hydrogels are wet and soft, researchers have found ways to laminate stretchable electronic skins onto hydrogels. Zhou et al. [2024] describe "sensor surface integration," where flexible strain gauge circuits are bonded to a hydrogel actuator's exterior. Because the hydrogel deforms, the attached sensor deforms in sync, capturing data like curvature or force. This approach can use mature flexible electronics technology, though maintaining adhesion between the hydrogel and sensor can be challenging (often, a thin intermediary layer or polymer coating is used to bond them).

DOI: http://dx.doi.org/10.5772/intechopen.1013161

The ultimate goal is a **closed-loop soft actuator** that senses its state and environment and adjusts accordingly – essentially imitating a muscle with built-in proprioception and touch sensing. Hydrogels are well-suited to this because their properties (conductivity, dielectric constant, etc.) respond linearly over a wide range of deformation, and because they can host ionic conduction (like biological tissues do).

A vivid demonstration of integrated sensing in hydrogel systems is a *self-sensing hydrogel gripper*. Zhang et al. (2023) [16] fabricated a bilayer hydrogel actuator with an intrinsically conductive polymer network. When the gripper bends to grasp an object, the internal resistance changes, which they measured to detect the gripping state – effectively, the gripper "feels" when it has an object and with how much force, without any external sensor [3]. This kind of built-in feedback is critical for delicate tasks like handling fragile objects or interacting with unpredictable environments.

Beyond proprioception, hydrogels can serve as *external sensors* integrated with actuators for environmental feedback. For example, a **hydrogel-based ionic pressure sensor** can be placed on a robotic fingertip (made of hydrogel or elastomer) to detect contact forces via changes in capacitance. Or a hydrogel that fluoresces in the presence of certain chemicals could be used as an *artificial olfactory sensor* on a soft robot that also moves. Wang et al. [3] highlight that hydrogel sensors have been demonstrated for a wide array of signals: pressure (ionic gels used in insoles for gait analysis), touch/strain (as mentioned), humidity (hydrogels that swell or change conductivity with water uptake), temperature (thermochromic hydrogels that change color or transparency), and even specific biochemical targets (glucose-sensing hydrogels for medical robots). This versatility means a soft robot built largely from hydrogels could have a kind of **universal sensor matrix** – its very body changes in measurable ways when conditions around it change.

Integrating all this requires addressing some challenges: Hydrogels are mostly water, so keeping long-term stable electrical connections is nontrivial (corrosion or hydrolysis can occur). Encapsulation, or the use of conductive polymer electrodes (like PEDOT:PSS coatings that remain stable in water), can mitigate this. Another issue is calibration – hydrogel sensor signals can drift if the gel slowly loses water or if temperature changes. Advanced hydrogels that are dehydrated (or form organogels with solvents that don't evaporate easily) are being developed to improve long-term stability [1].

In conclusion, hydrogels offer a unique pathway to **embedding sensory feedback directly into soft actuators**, fulfilling the vision of muscle-like actuators with muscle-like sensation. The ability of hydrogels to deform and transduce signals similarly to living tissue stands to greatly simplify the design of soft robotic systems: rather than separately engineered sensors and actuators, a single hydrogel component can play both roles. This integration is a major research trend, moving us closer to fully **autonomous and adaptive soft robots** that can sense and react to their environment in a tightly coupled manner (much like natural organisms). As Zhou et al. (2024) [17] emphasize, combining actuation and sensing in soft robots is key to achieving intelligent behavior – and hydrogels are among the most promising materials to realize this integration.

6. Advances in fabrication of hydrogel actuators

Creating complex hydrogel-based robots demands equally innovative fabrication techniques. Hydrogels present some fabrication challenges: they can be brittle or viscoelastic when handled, many are not easily machinable once formed, and traditional manufacturing methods (molding, casting) can be limiting in terms of geometric complexity. In response, researchers have developed a variety of **additive manufacturing (3D printing) methods and novel crosslinking techniques** tailored to hydrogels. These advances allow for the construction of intricate hydrogel architectures, multimaterial devices, and even the embedding of living cells for biohybrid actuators.

3D printing technologies: A major enabler in recent years is the adaptation of 3D printing to hydrogels (often termed *4D printing* when considering the shape change over time).

Figure 4 showcases several 3D printing techniques used to fabricate hydrogel actuators, each with its own strengths [7]:

- **Inkjet printing (Figure 5A):** Deposits tiny droplets of a liquid pre-polymer solution in a patterned way, which are then cured (typically by UV light or heat). Inkjet can create multi-material hydrogel patterns with high resolution in the plane. For example, researchers printed swelling and non-swelling regions in a hydrogel sheet via inkjet to create a self-folding origami structure that changes shape when hydrated [7]. Inkjet is relatively fast and good for thin, planar actuators, though building thick structures is more challenging due to layer-by-layer droplet deposition.

- **Two-photon polymerization (Figure 4B1-B2):** A laser-based method where a focused femtosecond laser induces polymerization in a tiny focal

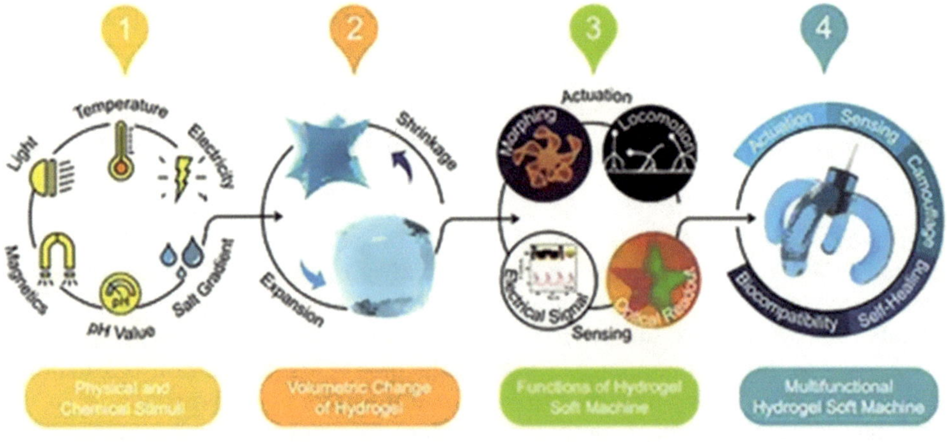

Figure 4.
Schematic illustration of a smart hydrogel-based soft actuator performing multiple functions: shape morphing actuation, locomotion, and self-sensing. Hydrogel devices can be designed to not only deform (actuate) in response to stimuli but also to produce an electrical signal or optical change corresponding to that deformation (proprioceptive sensing). This integration of actuation and sensing in the same soft material is a step toward intelligent, feedback-controlled soft robots [4, 18], locomotion [19], sensing with electrical signal output [20], and sensing with optical readout (ashley, 2003). copyright 2020, elsevier.

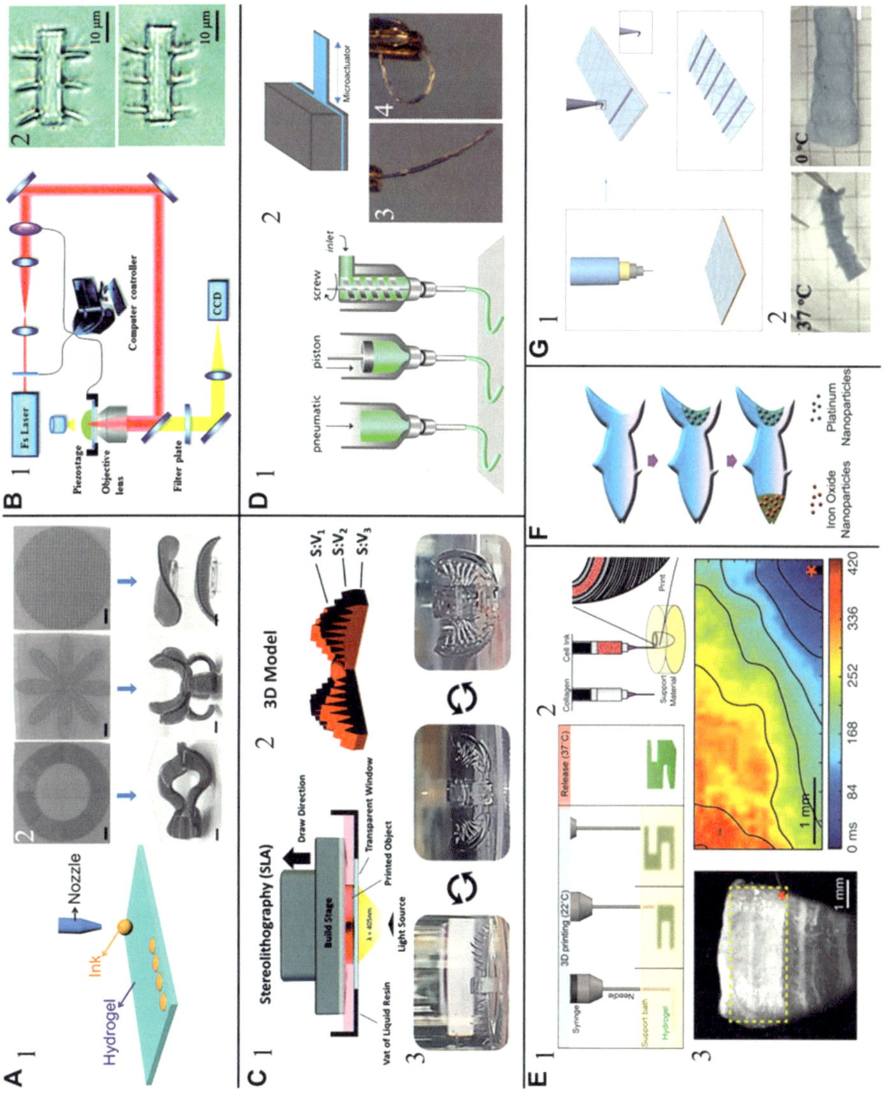

Figure 5.
Additive manufacturing strategies for hydrogel-based actuators. This includes extrusion printing, laser-based microfabrication, stereolithography (SLA), embedded printing, pneumatic/screw-driven deposition, bio-ink integration, and nanoparticle-enhanced printing approaches [7].

volume within a photosensitive hydrogel resin. By scanning this laser in 3D, arbitrary micro-structures can be "written" with sub-micron precision. Two-photon polymerization (2PP) can create extremely fine and complex hydrogel micro-actuators, such as micrograppers or microrobots that are smaller than a hair's width. The trade-off is that 2PP is slow for building anything beyond the millimeter scale and requires expensive laser setups. It excels at prototyping intricate micro-scale hydrogel devices (for instance, microneedle arrays or micro-swimmers with detailed features) [15].

- **Stereolithography (Figure 5C):** SLA uses a UV laser or projector to cure entire layers of a photopolymerizable hydrogel resin at once. Each layer is a cross-section of the 3D object, and layers are built up vertically. This method achieves good resolution and relatively rapid fabrication of centimeter-scale parts. SLA has been used to fabricate, for example, a bioinspired hydrogel fish tail with embedded stiff and soft regions to replicate fin motion. The limitation is that it usually requires the hydrogel resin to be somewhat modified for photo-curing, and that any unsupported overhangs need support structures (which must be removed later, carefully from a soft gel part).

- **Extrusion-based 3D printing (Figure 5D):** Also known as direct ink writing, this involves extruding a continuous filament of hydrogel precursor (or a gel ink) through a nozzle, depositing filaments that form the object. It's analogous to the way an FDM plastic printer works, but with gel. This technique is versatile and can print large structures. Researchers have printed hydrogel pneumatic networks, spring-like structures, and even entire soft grippers using extrusion printing [7]. A key advancement here is maintaining the shape fidelity of the printed filament – often **shear-thinning hydrogel inks** or **gelatin-supported printing** (printing into a supportive slurry that holds the filament until it solidifies) are used. One popular method is the **FRESH (Freeform Reversible Embedding of Suspended Hydrogels)** technique, where a hydrogel is printed into a bath of supportive gel that holds the print in place **(Figure 4E1-E3)**. This allows printing very soft hydrogels in complex shapes (like anatomically accurate organs or muscle-like structures) that would collapse if printed in air.

- **Mixed-mode and hybrid fabrication (Figure 5F, 5G):** Some actuators benefit from combining methods – for example, printing a hydrogel structure and then electrospinning fibers onto it, or printing a rigid frame and then casting a hydrogel around it (3D printing of molds). Sun et al. [7] illustrate a case (**Figure 5F**) where a magnetic filament was 3D printed within a hydrogel matrix – a mixed-mode approach yielding a magnetically responsive actuator. **Figure 5G** shows an example of a biobot where a 3D printed hydrogel scaffold is combined with living cells (engineered muscle cells), which contract to actuate the structure [16]. These hybrid strategies are powerful: they merge the best qualities of multiple materials (e.g., the strength of a polymer frame with the actuating ability of a hydrogel, or the contractile force of biological muscle with the patternable hydrogel skeleton).

Beyond 3D printing, **chemical crosslinking techniques** are crucial for controlling hydrogel properties during fabrication. Recent work has produced

tough hydrogels through multi-step crosslinking; for example, first forming a soft polymer network, then infusing a second monomer and polymerizing it in situ to form an interpenetrating double network. This approach, combined with micromolding or printing, yields actuators that don't tear easily, even under large strains [4]. Table 4 in [3] catalogs various crosslinking methods (thermal, UV, enzyme-mediated, click chemistry, etc.) and the literature that uses them to fabricate hydrogels for soft robots. The choice of crosslinking impacts not just mechanical strength but also biocompatibility (important if cells are involved) and stimulus responsiveness (e.g., a loosely crosslinked gel might swell more and deform more, but be weaker).

Integration of functional additives: Another fabrication consideration is how to incorporate nanoparticles, magnetic particles, or conductive elements uniformly into the hydrogel. For magnetic or optical actuation, particles must be well-dispersed. Techniques like in situ precipitation (forming magnetic nanoparticles directly inside the gel) or solvent exchange to load pre-formed particles are used. Ensuring homogeneous distribution is key to consistent actuation behavior [21].

Fabrication of composite structures: In many cases, a hydrogel actuator may be part of a composite – for instance, a **hydrogel-elastomer hybrid**. A common approach is to 3D print or cast a silicone rubber skeleton and then fill or coat it with hydrogel. One example is a soft gripper with a rubber exterior but a swelling hydrogel interior: when the hydrogel swells (via chemical stimulus), it pushes the rubber skin to curl the gripper fingers. Fabrication here requires bonding hydrogel to elastomer – achieved by a thin chemical primer that adheres to both or by curing them together in sequence so that interpenetration occurs at the interface [22].

In summary, advanced fabrication methods now allow the creation of **highly customized hydrogel actuators**: From tiny micro-robots made by laser-writing, to large-scale devices printed in support gels, to multi-material composites combining hydrogels with stiffer components or living cells. These tools are accelerating the development of soft machines with increasingly complex shapes and functionalities. We can now fabricate a soft robotic hand with hydraulically actuated hydrogel fingers, or a biohybrid robot with muscle cells on a printed hydrogel scaffold – things that were not feasible a decade ago. The continued improvement of fabrication techniques (higher resolution, faster printing, better bonding of dissimilar materials) will further expand design possibilities for soft robotic actuators and sensors.

7. Applications of hydrogel-based soft actuators

Hydrogel actuators and artificial muscles are finding applications across a broad range of fields, thanks to their unique combination of softness, wetness, and responsiveness. Here, we highlight several major application domains and examples:

- **Biomedical devices and healthcare:** This is arguably the most naturally aligned area for hydrogel actuators. Their biocompatibility and gentle mechanics make them ideal for use *in vivo* or in contact with biological tissue. One exciting application is **targeted drug delivery**. For instance, magnetically controlled hydrogel micro-robots have been developed that can navigate bodily fluids and deliver drugs to specific locations (e.g., a hydrogel loaded with drugs that swells or releases its payload when triggered by a magnetic field or by reaching a target site) [3]. Another example is **soft hydrogel grippers for minimally invasive surgery**, which can deploy through a small incision and then swell to grasp organs or tissues delicately without causing damage. In the realm of wearables, hydrogel-based artificial muscles have been proposed for assistive physiotherapy devices – e.g., a sleeve with hydrogel actuators that contract to massage muscles or help blood circulation. The transparent and soft nature of hydrogels is also leveraged in *implantable devices*: Researchers have made hydrogel-based cardiac patches that mechanically actuate (either via electrical or optical stimulation) to assist heart contraction, essentially a soft actuator to help a weakened heart pump blood. Because the hydrogel matches the mechanical properties of heart tissue and is mostly water, it integrates without causing inflammation [3]. Another medical use is dynamic wound dressings – a hydrogel that can apply mechanical stimulation to a wound (promoting healing) or can close/open in response to wound conditions. Overall, in biomedical applications, the emphasis is on **safety, precision, and biocompatibility**, where hydrogels excel even if high force output is not required.

- **Soft robotics and manipulation:** Hydrogel actuators are used to make soft robotic grippers and manipulators, especially for delicate tasks. A noteworthy example is a **hydrogel gripper that caught a live goldfish without harm** [7]. This gripper's fingers were made of a very low-stiffness hydrogel, rendering it almost invisible and imparting a very gentle touch (the fish was not injured, demonstrating how soft the grasp was). The gripper could lift and release the fish using minimal force – something rigid grippers or even silicone grippers might struggle with due to slippage or pressure points. In food handling or biomedical sample handling, hydrogel fingers could safely manipulate fragile, slippery objects (like soft fruits, tissues, or eggs) with reduced risk of contamination (being water-based, they don't shed particles). Soft hydrogel robots have also been demonstrated that *crawl or walk*: for instance, an inchworm-like robot made of a PNIPAM hydrogel strip that oscillates under cyclic heating and cooling, achieving locomotion. Although slow, these robots show that completely soft, even transparent robots are possible, which could be useful for tasks like moving in confined wet environments (maybe inside lab-on-chip devices or even within the human gastrointestinal tract for diagnostics) [14].

- **Wearable and assistive devices:** Hydrogels' flexibility and skin-like feel make them attractive for wearable robotics and haptic devices. We mentioned hydrogel strain sensors for wearables; when combined with hydrogel actuators, one can create devices like a soft exosuit sleeve that both senses muscle signals and assists movement [10]. Hydrogels have been explored for **artificial skin with haptic feedback** – e.g., a patch that actuates to provide a sensation to the user. Because hydrogels can be made to change stiffness (by controlling water content or by polymerizing additional crosslinks on demand), a wearable hydrogel could even simulate varying muscle tone or massage patterns on the body. Another interesting concept is **assistive gloves**: A glove with hydrogel pneumatic actuators or electrically driven hydrogel muscles can help stroke patients open and close their hands gently. Unlike typical assistive gloves made of silicones (which, while soft, are still dry and can cause sweat and discomfort), a hydrogel glove could, in theory, keep the skin hydrated and comfortable during use.

- **Environmental and agricultural robotics:** In agriculture, hydrogels are already used for water retention in soil. Coupling this with robotics, one could create soft hydrogel-based systems that respond to moisture levels – essentially *robotic soil conditioners* or slow-release irrigation actuators. For example, a hydrogel that expands when soil is dry (pushing water out) and contracts when soil is moist could act as an autonomous irrigation valve. Liu et al. [23] demonstrated highly absorbent hydrogel constructs in farmlands that regulate fertilizer release and water delivery to plants, functioning like a smart substrate [3]. Additionally, hydrogel sensors can monitor environmental conditions (humidity, pollutants) and actuate vents or signal changes. The biocompatibility is a bonus here, as hydrogels used in a field will eventually biodegrade. In environmental cleanup, hydrogel soft robots (for instance, a soft hydrogel blob that can deform and navigate through cracks) could potentially be used to explore or remediate environments – a concept in its early stages but intriguing [24].

- **Food and consumer products:** Hydrogels are actually common in food (think gelatin desserts or hydrocolloids in cooking). Edible hydrogel robots have been proposed – for instance, an edible hydrogel pill that can move in the stomach to aid in mixing or sensing conditions, then dissolves. In consumer products, one might see hydrogel-based haptic feedback devices (since they can be very comfortable on skin). Also, toys or educational robots made of hydrogels could be safe for children (soft, no sharp edges, even ingestible if truly edible).

- **Emerging high-tech applications:** As an emerging frontier, researchers are exploring **optically transparent robots** using hydrogels. For instance, a nearly invisible soft actuator that can camouflage in water or in certain environments has military or research applications (such as observing marine life without disturbing them). Hydrogels can be made transparent, so a robot arm in an aquarium could be practically invisible while manipulating objects. Additionally, in microfluidics and lab-on-chip systems, hydrogel actuators (like microvalves, pumps, and micrograppers) are being integrated to handle biological samples gently and precisely [3].

It is worth noting that many of these applications are still in prototype or proof-of-concept phases. Hydrogel robots have shone in demonstrations like the fish-grabbing gripper or tiny drug-delivery crawlers, but scaling them up to practical, everyday use involves addressing durability, control, and manufacturing yield issues. Nonetheless, the range of applications already explored shows the **versatility of hydrogel-based actuators** – from healthcare to food handling to environmental monitoring, their unique properties open avenues that conventional robots cannot easily enter.

7.1 Recent advances in smart hydrogel-based actuators and artificial muscles

In the past few years, the field of smart hydrogel actuators has progressed rapidly, with researchers focusing on strategies to overcome limitations in strength, responsiveness, and durability, while broadening application domains.

7.1.1 Material innovations and nanocomposites

Recent studies highlight the importance of reinforcing hydrogel matrices with nanomaterials to enhance stress output and longevity. 4, reviewed advances in nanocomposite hydrogels, emphasized how fillers such as graphene and MXene can provide conductivity and mechanical strength without compromising softness. 1, reported that double-network hydrogels and fiber-reinforced designs significantly improved load-bearing capacity and resistance to fracture, making them suitable for repetitive actuation cycles in artificial muscle applications.

7.1.2 Additive and 4D printing strategies

Additive manufacturing remains central to tailoring hydrogel architectures. Sun et al. [7] demonstrated how stereolithography and extrusion printing can create patterned, multi-material hydrogel actuators. Building on this, [22] and [25] showed the integration of multi-material deposition for complex robotic functions, while [26] highlighted how hybrid strategies combining extrusion with embedded printing yield actuators with spatiotemporal programmability. More recently, [3] extended these ideas into the domain of 4D printing, reporting hydrogel continuum robots capable of adaptive bending and multimodal locomotion under combined stimuli.

7.1.3 Artificial muscle development

The transition of hydrogels from simple actuators to functional artificial muscles is becoming more feasible. Jing et al. [2] emphasized the evolution of hydrogel-based artificial muscles, with notable progress in achieving faster contraction–relaxation cycles, greater strain, and improved fatigue resistance. López-Díaz et al. [1] similarly noted that integrating multiple stimuli-responsiveness (e.g., thermal and electrical) allows hydrogels to act more like biological muscles, capable of precise control and adaptive movement.

7.1.4 Biohybrid systems and biomedical applications

Biohybrid hydrogels have also emerged as a frontier for integrating living cells with synthetic materials. Recent demonstrations, reported by [1] and [2], show that embedding cardiomyocytes in hydrogel scaffolds enables spontaneous rhythmic contractions, providing a platform for next-generation bio-actuators. In parallel, [3] described hydrogel microrobots powered by magnetic nanoparticles, achieving multimodal locomotion, such as rolling and jumping, with clear implications for targeted drug delivery and minimally invasive biomedical devices.

Together, these recent advances highlight how hydrogel actuators are evolving from conceptual models into robust artificial muscles and microrobotic systems. By combining nanocomposite reinforcement, programmable fabrication, and biohybrid integration, researchers are moving closer to realizing intelligent, multifunctional soft robotic systems capable of performing in real-world environments.

8. Emerging frontiers and research directions

Figure 6 provides a schematic overview of the hierarchical development pathway of smart hydrogel-based actuators, from material innovations to system-level applications.

The field of hydrogel-based soft robotics is rapidly evolving, and several exciting frontiers are emerging:

- **Multi-responsive and programmable hydrogels:** Next-generation hydrogels are being designed with *multiple responsiveness*. For instance, a single hydrogel that reacts to temperature, pH, and electric fields in different ways, or different regions of a composite hydrogel, each tuned to a specific stimulus. Sun et al. [7] reported a multi-responsive hydrogel actuator that combined thermal sensitivity (via PNIPAM), magnetic sensitivity (via

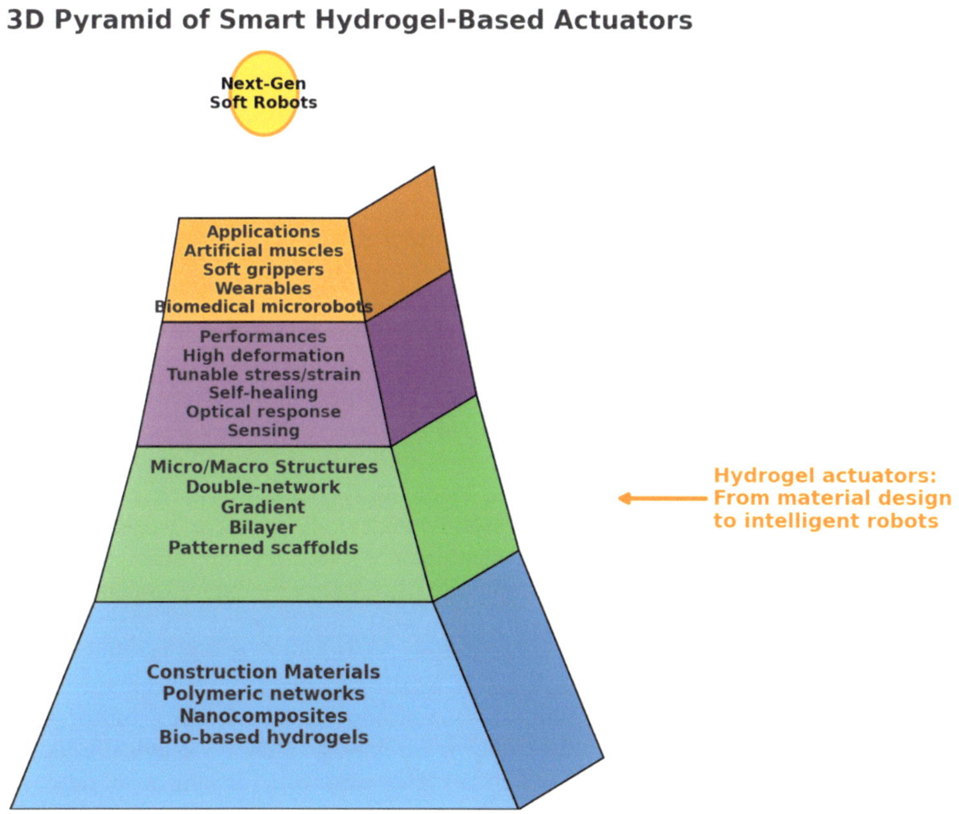

Figure 6.
3D pyramid schematic illustrating the development pathway of smart hydrogel-based actuators.
*The pyramid highlights how advances in **construction materials** (e.g., polymeric networks, nanocomposites, bio-based hydrogels) enable **micro/macro structural innovations** (double-network, gradient, bilayer, patterned scaffolds). These structures give rise to enhanced **performances** such as high deformation, tunable stress/strain, self-healing, optical response, and proprioceptive sensing. At the top, these advances culminate in **applications** including artificial muscles, soft robotic grippers, wearable exosuits, and biomedical microrobots. The **golden sphere on the apex** symbolizes the ultimate vision:* next-generation intelligent soft robots that integrate actuation, sensing, and adaptability *[1, 3, 4].*

embedded nanoparticles), and even solvent sensitivity, enabling it to be controlled by either stimulus or to perform complex tasks when stimuli are applied in sequence [3]. Along with this, **programmable shape change** is a major goal: using techniques like 4D printing to pattern the anisotropy or swelling behavior in the hydrogel, researchers can "program" how a hydrogel will morph when activated. This means a single piece of hydrogel can be encoded to transform into a predetermined 3D shape (like an origami structure or a functional device) when a stimulus is applied – effectively embedding an algorithm into the material's structure. Such programmed hydrogel actuators could be shipped flat/dry and then activated on-site (by hydration or heating) to form the desired soft robot [27].

- **Integration of living cells (biohybrid robots):** An especially intriguing frontier is merging hydrogels with biology. Hydrogels can serve as scaffolds to culture living muscle cells (myocytes) or other contractile cells. These living cells then act as actuators by contracting, with the hydrogel providing structure and perhaps additional actuation of its own. Researchers have created small "bio-bots" where neonatal cardiac muscle cells were grown on a hydrogel frame; the spontaneous contraction of the heart cells made the frame twitch and even walk in a legged design (when the cells contract in sequence like legs) [7]. Similarly, skeletal muscle cells have been cultured on hydrogel ribbons to make controllable biohybrid actuators that respond to electrical stimulation (the cells will contract when electrically pulsed). The hydrogel provides a tissue-like environment for the cells and can be designed to transmit the force of cell contraction to perform work. In the future, **biohybrid artificial muscles** might combine the strength and efficiency of natural muscle tissue with the resilience and programmability of synthetic hydrogels. This field also connects to tissue engineering and regenerative medicine – the artificial muscles being developed could potentially be implanted to assist or replace damaged muscles in patients.

- **Tough and self-healing hydrogels:** Mechanical robustness is a critical area of improvement. New hydrogel chemistries, such as slide-ring hydrogels, double/triple-network hydrogels, and nanocomposite hydrogels, are dramatically improving tear resistance and fatigue life [4]. We are seeing hydrogels that can be stretched 1000% and not break, or sustain millions of cycles of deformation. Incorporating these into actuators means the devices will last longer and handle higher loads. Additionally, **self-healing hydrogels** that can autonomously repair small cuts or punctures are emerging. A hydrogel actuator made of self-healing material could recover from damage (like a biological muscle that heals after micro-tears), which is highly desirable for long-term deployment or for robots operating in harsh, unpredictable environments. Some self-healing hydrogels use dynamic crosslinks (like ionic bonds or host-guest interactions) that can break and reform. If a crack forms, the polymer chains can reattach across it, given some time or a stimulus (like pH or heating) to trigger healing. Imagine a soft gripper that, if cut by a sharp object, could be "healed" by simply dipping in a saline solution for a few hours to re-form its network bonds [28].

- **Out-of-water operation and longevity:** Historically, many hydrogel actuators needed to remain in aqueous environments to function (to avoid drying out and to leverage fluid interaction, e.g., ionic actuators in electrolyte). A current frontier is making hydrogels that can operate **in air for extended periods** without desiccation. Strategies include developing *organo-hydrogels* (hydrogels infused with organic solvent or oil that doesn't evaporate easily), or hydrogels with protective coatings that retain moisture. López-Díaz et al. [1] added graphene to a hydrogel to act as a thermal filler that dissipates heat and reduces water evaporation, enabling an ionic hydrogel actuator to bend repeatedly in air for a longer time before drying out. Another approach is to encapsulate the hydrogel in a thin elastomer membrane – the membrane keeps water in but is flexible enough not to impede motion significantly. These advances will allow hydrogel robots to venture outside the lab and operate in everyday air environments, greatly expanding their usability (e.g., a hydrogel-based artificial muscle that could work in open air for days, powering a soft exoskeleton).

- **Scaling up and scaling down:** Two divergent but important directions are scaling hydrogels to larger and smaller sizes. On the large end, can we make meter-scale hydrogel actuators? One challenge is that diffusion is slow in large hydrogels, so response times become very long. Researchers are looking at embedding vascular networks or porous structures in big hydrogels to facilitate rapid stimulus delivery throughout the material. This could enable large hydrogel muscles that actuate more quickly. On the small end, micro- and nano-scale hydrogel robots (sub-millimeter size) are very attractive for biomedical use (imagine a swarm of microscopic soft robots cleaning arteries or delivering drugs). Two-photon polymerization and DNA self-assembly techniques are being developed to create nanoscale hydrogel structures. Also, **DNA hydrogels**, which are networks made of DNA strands, can function at the nanoscale and be programmed at a molecular level for specific responses – a very futuristic concept where a robot could be the size of a cell, made of a biocompatible hydrogel that responds to biochemical signals [29, 30].

- **Intelligent control and modeling:** As hydrogel actuators become more complex, the field is also pushing forward in terms of how to model and control them. Hydrogels involve coupled chemical-mechanical dynamics (e.g., diffusion of ions and elastic deformation), which are quite different from traditional motor control. New computational models and simulation tools are being developed to predict hydrogel robot behavior under various stimuli [Zhou et al., 2024]. Machine learning is even being applied to learn the inverse model of a hydrogel actuator (i.e., how to drive it to achieve a desired motion). We might see AI-controlled hydrogel robots that can adaptively tune their stimulus inputs (like voltage or light intensity) to account for non-linear responses or environmental changes (like a slight temperature shift affecting swelling). This kind of high-level intelligence, paired with the inherent smart material, could result in **highly adaptive soft robotic systems**.

Overall, the emerging frontiers in hydrogel-based soft robotics are moving toward **more life-like systems** – in material composition, functionality, and autonomy. We foresee robots that are literally *built like living creatures*, with soft hydrogel muscles and skins, sensorimotor feedback loops within their material, the ability to heal and adapt, and perhaps even integration with living cells. Hydrogel actuators are at the core of this vision, as they blur the line between machine and organism with their tissue-mimicking properties. As research continues, we anticipate hydrogel artificial muscles performing tasks in environments and ways that were previously unimagined for robots, from inside the human body to the depths of the ocean, all while being safe and environmentally benign.

9. Challenges and future outlook

Despite the tremendous progress and exciting potential of hydrogel-based actuators, several significant challenges remain before they achieve widespread practical adoption:

- **Mechanical limitations: Low strength and slow response** are the Achilles' heel of many hydrogels. While innovations like double-network hydrogels have improved toughness, most hydrogels cannot yet match the stress outputs of pneumatic or shape-memory actuators. The intrinsic viscoelasticity means they can be sluggish – large hydrogel pieces might take minutes to fully actuate. Future work must continue to improve mechanical properties (through material chemistry and composite design) and find clever ways to speed up responses (like incorporating porosity or mini-channels for faster stimuli diffusion). One promising direction is the use of **porous or nanostructured hydrogels** that allow rapid water transport, thus speeding swelling/deswelling cycles. Another is combining hydrogels with fibers or textiles to bear load (the hydrogel provides motion, the fiber provides strength). By addressing these, we could see hydrogel muscles that contract in milliseconds and lift kilograms, truly competing with electromagnetic motors or pneumatic cylinders in some applications.

- **Dehydration and environmental stability:** Operating in ambient conditions introduces the issue of water loss. A dried-out hydrogel is essentially useless as an actuator. Even with coatings or organogels, maintaining long-term hydration is challenging. This limits shelf-life and operational duration. Devices might need periodic "recharging" with water or storage in humid environments. Solving this could involve developing *hydrogels that self-regulate their hydration* (for example, hydrogels that can absorb moisture from the air when they start drying) or integrating microfluidic systems that can replenish water on the go. Also, environmental factors like temperature or pH can unintentionally alter hydrogel performance – robustness against such fluctuations is needed for reliable robots. This might be tackled by new hydrogel formulations that are less sensitive to ambient changes except for the intended stimulus [31].

- **Fabrication and scalability:** While we highlighted advanced fabrication methods, scaling those up for mass production is non-trivial. 3D printing is still relatively slow for large volumes; for commercial products, faster manufacturing methods (roll-to-roll printing, injection molding of hydrogel precursors, etc.) may be required. However, molding hydrogels can be tricky, as they can stick or tear. Achieving reproducibility in hydrogel actuator properties is another challenge – slight variations in crosslink density or ingredient mixing can lead to different swelling behavior. Quality control will be important. New techniques, like **embedded 3D printing** and continuous manufacturing of hydrogel fibers, might help produce standardized artificial muscles in bulk.

- **Control and hysteresis:** Hydrogels often exhibit hysteresis (not returning exactly to the original state due to slow relaxation or incomplete swelling). Precise control of position or force can be more challenging than with traditional actuators. Developing robust control algorithms that can handle the non-linear, history-dependent behavior of hydrogels remains an ongoing challenge. Additionally, most hydrogel actuators are still open-loop (except for those with integrated sensing); implementing closed-loop control requires reliable sensor feedback and possibly predictive models. Progress in computational modeling (as mentioned in Frontiers) is critical to enable model-based control of soft hydrogel systems.

- **Integration with electronics:** If a hydrogel robot requires electronics (for power or communication), integrating those rigid components is tough. How do you attach a circuit or battery to a jelly-like actuator without compromising its softness or drying it out? Efforts are underway to create **soft batteries and circuits** (using, for example, conductive hydrogels as wires or encapsulating electronics in stretchable, waterproof membranes). For now, many prototypes keep electronics external (wires leading to the hydrogel). Fully untethered hydrogel robots will need innovative solutions like flexible printed circuit boards that can gel with the hydrogel or wireless energy transfer (so no on-board battery is needed). This is an interdisciplinary challenge bridging materials, electrical engineering, and robotics [9].

- **Longevity and fatigue:** Repeated swelling can eventually wear out a hydrogel (mechanically or chemically). The polymer chains might break down over time, especially under UV light or if there are biological agents (for in vivo applications). Biocompatible hydrogels might be biodegradable, which is great for not leaving waste, but it means the actuator may degrade after accomplishing its task (like a drug-delivery bot that dissolves). If longevity is needed, non-degrading hydrogels or protective measures must be in place. Fatigue testing of hydrogel actuators is needed to understand how many cycles they can undergo – currently, many prototypes operate for maybe hundreds of cycles; getting to millions (like a car engine or an industrial robot requires) is a big challenge.

- **Safety and regulatory hurdles:** For medical applications, hydrogels are typically biocompatible, but any additional components (nanoparticles, crosslinking agents) must be non-toxic and must not elicit adverse immune responses.

Rigorous testing and regulatory approval will be needed before hydrogel robotic devices can enter clinical use. Similarly, the environmental deployment of hydrogels means considering ecological impact – while many hydrogels are benign, additives like nanomaterials could pose risks if they leach out. Developing **eco-friendly hydrogel systems** that either biodegrade into harmless components or can be recollected is important for sustainable design [30, 32].

Despite these challenges, the overall outlook for smart hydrogel actuators is highly optimistic. The pace of research is rapid, and interdisciplinary efforts are converging to address the shortcomings. **Materials scientists** are inventing stronger, smarter hydrogels; **engineers** are creating new fabrication tools and device designs; **roboticists** are learning how to model and control these squishy mechanisms, and **biologists/chemists** are integrating living systems and novel chemistries. The challenges listed are actively being tackled:

- Projects like using zwitterionic or hydrophobic modifications to reduce hydrogel evaporation show promise for addressing the dehydration issue.

- There are first demonstrations of self-healing actuators that regain function after being cut, addressing longevity.

- For control, researchers like Zhou et al. have proposed embedding multiple sensors and using machine learning to estimate the state of a hydrogel actuator in real-time, which could compensate for hysteresis and uncertainty [Zhou et al., 2024].

In the long term, success in this field could revolutionize how we design robots. Instead of rigid skeletons with discrete motors, we might literally "grow" robots out of polymer gels that move with life-like grace. Such robots could seamlessly interact with natural organisms (for medical interventions or environmental blending) without causing harm. We may see *soft exoskeletons* that behave as an extension of the human body, or artificial organs that can actively pump, contract, and sense just like real organs. The concept of **"living machines"** – robots that share traits with living systems – is brought closer by hydrogel technology, since hydrogels themselves straddle the boundary between inorganic material and living tissue.

In conclusion, smart hydrogel-based actuators and artificial muscles represent a cutting-edge frontier in soft robotics, one that holds immense promise for creating robots that are safer, more adaptive, and more integrated with the living world. While significant technical challenges remain, ongoing research is rapidly chipping away at these barriers. The convergence of advances in materials (e.g., multi-responsive and tougher hydrogels), fabrication (4D printing and biofabrication), and control (sensor integration and AI) suggests that the coming years will witness hydrogel soft robots transitioning from laboratory curiosities to real-world solutions. By harnessing the qualities of hydrogels, we edge closer to robots that truly emulate the elegance of biological systems – a transformative vision for the future of robotics.

10. Conclusion

Hydrogels bring a unique arsenal of properties to soft robotics – softness, wetness, stimuli-responsiveness, and biocompatibility – enabling new kinds of actuators and artificial muscles that traditional materials cannot achieve. The chapter has detailed the fundamentals of hydrogel actuation, surveyed actuation mechanisms and their use as artificial muscles, explored how sensing can be inherently combined, and reviewed advanced fabrication techniques and applications. The emerging research frontiers show that, while challenges exist, they are surmountable with creative interdisciplinary efforts. In a broader perspective, smart hydrogel actuators are more than just components for soft robots; they exemplify a shift toward **material-level intelligence**, where the boundary between structure and function blurs. As this field matures, we anticipate that hydrogel-based soft robots will play pivotal roles in medicine, caregiving, environmental stewardship, and beyond – functioning in harmony with living systems and pushing the boundaries of what robots can do and where they can go. The ongoing journey of overcoming challenges and refining these devices is not only technologically significant but also deepens the inspiration drawn from biology, ultimately bringing robotic systems closer to living organisms in capability and form. With continued innovation, the vision of robust, agile, and perceptive "gel robots" – once a mere science fiction concept – is steadily becoming a reality, heralding a new era of **truly soft robotics**.

Conflict of Interest

The author(s) declare no conflict of interest.

Author details

Nazia Mustary Nayna[1*] and Sabrina Sabnam Samonty[2]

1 School of Mechanical and Mechatronic Engineering, University of Technology Sydney (UTS), Sydney, Australia

2 Islamic Bank Medical College, Rajshahi, Bangladesh

*Address all correspondence to: naziamustary528@gmail.com

References

[1] López-Díaz D, et al. Hydrogels in Soft Robotics: Past, Present, and Future. 2024

[2] Jing Z, et al. Advances in Artificial Muscles: A Brief Literature and Patent Review. 2023

[3] Wang Y, et al. Hydrogel-Based Continuum Soft Robots: Progress and Prospects. 2025

[4] Goyal R, Mitra S. A brief overview of bioinspired robust hydrogel based shape morphing functional structure for biomedical soft robotics. Frontiers in Materials. 2022;**9**:837923

[5] Zhao Q, Yu Z, Liang Y, Ren L, Ren L. Intelligent hydrogel actuators with controllable deformations and move ments. Frontiers in Materials, 2021;**8**: 661104. DOI: 10.3389/fmats.2021.661104

[6] Zhang Y, Liu X, Wang Y, Chen H, Li J. Rational Design of Smart Hydrogels for Biomedical Applications. Front Chem. 2021;**8**:615665. DOI: 10.3389/fchem.2020.615665

[7] Sun J, et al. 3D Printing Hydrogel-Based Soft and Biohybrid Actuators. 2021

[8] Cheng H, L Long, J Cao, S Zhang, Y Wang. Dual-crosslinked mussel-inspired smart hydrogels with enhanced antibacterial and angiogenic properties for chronic infected diabetic wound treatment via pH-responsive quick cargo release. Chemical Engineering Journal. 2021;**411**:128564. DOI: 10.1016/j.cej.2021.128564.

[9] Li M, Wang X, Dong B, Sitti M. In-air fast response and high speed jumping and rolling of a light-driven hydrogel actuator. Nature Communications. 2020;**11**:3988

[10] Rus D, Tolley M. Design, Fabrication and Control of Soft Robots, 2015

[11] Keplinger C, Sun JY, Foo CC, Rothemund P, Whitesides GM, Suo Z. Stretchable, transparent, ionic conductors. Science. 2013;**341**: 984–987. DOI: 10.1126/science.1240228

[12] Majidi C. Soft Robotics: A Perspective—Current Trends and Prospects for the Future. 2014

[13] Hines L, Petersen K, Lum GZ, Sitti M. Soft actuators for small-scale robotics. Advanced Materials. 2017;**29** (13):1603483. DOI: 10.1002/adma.201603483

[14] Trivedi D, et al. Soft Robotics: Biological Inspiration, State of the Art, and Future Research. 2008

[15] Polygerinos P, Correll N, Morin SA, Mosadegh B, Onal CD, Petersen K, Cianchetti M, Tolley MT, Shepherd RF. Soft robotics: Review of fluid-driven intrinsically soft devices; manufacturing, sensing, control, and applications in human-robot interaction. Advanced Engineering Materials. 2017;**19**(12):1700016. DOI: 10.1002/adem.201700016

[16] Zhang S, Wei H, Tang J. Self-sensing magnetic actuators of bilayer hydrogels. International Journal of Smart and Nano Materials, 2023;**14** (4):496–509. DOI: 10.1080/19475411.2023.2257616

[17] Zhou X, Yu X, You T, Zhao B, Dong L, Huang C, Zhou X, Xing M, Qian W, Luo G. 3D Printing-Based Hydrogel Dressings for Wound Healing. Adv. Sci. 2024;**11**:2404580. DOI: 10.1002/advs.202404580

[18] Fu F, Shang L, Chen Z, Yu Y, Zhao Y. Bioinspired living structural color hydrogels. Sci Robot. 2018 Mar 28;**3**(16):eaar8580. DOI: 10.1126/scirobotics.aar8580.

[19] DeForest CA, Anseth KS. Advances in bioactive hydrogels to probe and direct cell fate. Annu Rev Chem Biomol Eng. 2012;**3**:421–44. DOI: 10.1146/annurev-chembioeng-062011-080945.

[20] Humberto Lobato-Morales, Alonso Corona-Chávez, José Luis Olvera-Cervantes, Ricardo Arturo Chávez-Pérez, José Luis Medina-Monroy. Wireless Sensing of Complex Dielectric Permittivity of Liquids Based on the RFID. EEE Transactions on Microwave Theory and Techniques. 2014;**62**(9):2160–2167

[21] Ouyang Y, Huang G, Cui J, Zhu H, Yan G, Mei Y. Advances and challenges of hydrogel materials for robotic and sensing applications. Chemistry of Materials. 2022;**34**(21):9307–9328. DOI: 10.1021/acs.chemmater.2c01960

[22] Wallin TJ, Pikul J, Shepherd RF. 3D printing of soft robotic systems. Nature Reviews Materials. 2018;**3**:84–100. DOI: 10.1038/s41578-018-0002-2

[23] Liu H, Jia X, Liu R, Chen K, Wang Z, Lyu T, Cui X, Zhao Y, Tian Y. Multifunctional gradient hydrogel with ultrafast thermo-responsive actuation and ultrahigh conductivity. Journal of Materials Chemistry A. 2022;**10**:21874–21883. DOI: 10.1039/D2TA05770K

[24] Jiang J, Xu S, Ma H, Li C, Huang Z. Photoresponsive hydrogel-based soft robot: A review. Materials Today Bio. 2023;**20**:100657. DOI: 10.1016/j.mtbio.2023.100657

[25] Gul JZ, Sajid M, Rehman MM, Siddiqui GU, Shah I, Kim KH, . . . Choi KH. 3D printing for soft robotics – A review. Science and Technology of Advanced Materials. 2018;**19**(1):243–262. DOI: 10.1080/14686996.2018.1431862

[26] Stano G, Percoco G. Additive manufacturing aimed to soft robots fabrication: A review. Extreme Mechanics Letters. 2021;**42**:101079. DOI: 10.1016/j.eml.2020.101079

[27] Yuk H, Lin S, Ma C, Takaffoli M, Fang NX, Zhao X. Hydraulic hydrogel actuators and robots optically and sonically camouflaged in water. Nature Communications. 2017;**8**:14230. DOI: 10.1038/ncomms14230

[28] Shintake J, Cacucciolo V, Floreano D, Shea H. Soft robotic grippers. Advanced Materials. 2018;**30**(29):1707035. DOI: 10.1002/adma.201707035

[29] Kim S, et al. Soft Robotics: A Bioinspired Evolution in Robotics. 2013

[30] Calvert P. Hydrogels for soft machines. Advanced Materials. 2009;**21**(7):743–756. DOI: 10.1002/adma.200800534

[31] Whitesides GM. Soft robotics. Angewandte Chemie International

Edition. 2018;**57**(16):4258–4273. DOI: 10.1002/anie.201800907

[32] Manti M, Cacucciolo V, Cianchetti M. Stiffening in soft robotics: A review of the state of the art. IEEE Robotics and Automation Magazine. 2016;**23**(3):93–106. DOI: 10.1109/MRA.2016.2582718

Chapter 5

Hydrogel Engineering Interventions for Precision Cartilage Regeneration: Bridging Pathological Mechanisms and Therapeutic Innovation

Zhi Zheng, Roujun Chen and Yaqin Lin

Abstract

Although microfracture, osteochondral transplantation, and autologous chondrocyte implantation are commonly employed, accurately replicating the natural calcified layer at the cartilage-bone interface remains challenging, leading to structural and functional impairments in regenerated cartilage. For successful cartilage regeneration, engineered therapeutic materials must be tailored to the structural and pathophysiological characteristics of cartilage injury. This necessitates the development of strategic interventions based on a comprehensive understanding of the injured structure and microenvironment, including extracellular matrix degradation, inflammatory cascades, disrupted intercellular communication, and integrated pathophysiological networks. This review systematically elucidates microenvironment evolution and molecular mechanisms post-cartilage injury while highlighting the application of injectable hydrogels and 3D printing technology as delivery vehicles for various therapeutic agents (including drugs, cells, and genes) in tissue engineering. By emphasizing their unique capacity for designing customized systems to modulate the regeneration process, this work aims to establish a cohesive theoretical framework and a clear technical roadmap for guiding the future development of precise cartilage repair solutions.

Keywords: cartilage injury, pathological mechanism, injectable hydrogel, 3D printing, cartilage regeneration

1. Introduction

The osteochondral unit is a core functional structure located at the articular junction. It comprises a cartilage layer, a subchondral bone layer, and an intermediate calcified cartilage layer between them [1]. This unit possesses a unique hierarchical architecture and specific biomechanical properties. These characteristics are crucial for shock absorption, mechanical load-bearing, and smooth joint movement. However, cartilage tissue has a very limited capacity for self-repair because it lacks blood vessels, nerves, and lymphatic vessels. Once injured by trauma or degenerative disease, the damaged area cannot heal spontaneously and often progresses into an osteochondral defect [2]. This condition causes persistent joint pain, restricted mobility, and, in severe cases, a complete loss of joint function. Consequently, it significantly impacts patients' daily activities and quality of life.

The incidence of joint injuries, such as osteoarthritis and rheumatoid arthritis, is continuously increasing. This trend is driven by an aging population and greater public participation in sports activities [3]. Global statistics from 2020 estimated that around 595 million people suffered from joint injuries [4]. The resulting pain and treatment costs place a substantial economic burden on families and society. Therefore, developing effective therapies that can achieve fundamental cartilage repair and reverse the disease process is a critical challenge in orthopedics and sports medicine.

Current clinical management for osteoarthritis establishes a multi-tiered system, ranging from basic care to surgical replacement, yet remains primarily palliative. These strategies focus on symptom relief and delaying progression rather than reversing the underlying degenerative pathology or achieving long-term cartilage preservation and joint function. Standard approaches encompass pharmacological and surgical interventions. Pharmacological therapies often merely slow disease progression and can introduce side effects, including gastrointestinal injury or cardiovascular risks with prolonged use [5]. Surgical options, such as arthroscopic debridement, do not facilitate cartilage regeneration. Microfracture frequently yields unsatisfactory long-term results, while osteochondral autograft transplantation is constrained by limited graft availability and procedural complexity. Autologous chondrocyte implantation necessitates two separate surgeries and is costly, and allograft transplantation confronts challenges regarding graft preservation and safety [6–9]. A fundamental limitation common to all these methods is their inability to accurately reconstruct a regenerative microenvironment at the defect site that can concurrently modulate inflammation and promote the synthesis of a functional ECM.

To overcome these constraints, tissue engineering strategies have emerged that focus on constructing biomaterial scaffolds which mimic the native cartilage microenvironment [10, 11]. Among these, injectable hydrogels have garnered significant interest due to their unique advantages. Firstly, they provide a minimally invasive delivery platform that can be precisely implanted via injection to perfectly adapt to irregular defect geometries, thereby greatly reducing surgical trauma and associated complication risks [12]. More importantly, by serving as both a biomimetic and functional scaffold, they can replicate the physicochemical properties of the natural ECM to provide a supportive habitat for cells while also acting as versatile carriers [13]. Through the controlled delivery of therapeutic agents such as drugs, cells, or genes, they enable active and precise regulation of the injury microenvironment to guide the regeneration of functional hyaline cartilage [14, 15].

For more complex and large osteochondral defects, simple injectable hydrogels often face challenges related to their mechanical strength and structural biomimicry. In this context, three-dimensional (3D) printing technology serves as a complementary and advanced approach that offers the possibility of precise structural imitation. This technique facilitates the fabrication of scaffolds with biomimetic gradient structures, such as bilayer or multilayer cartilage-bone constructs, using hydrogels as bioinks [16, 17]. By enabling the spatial localization of different cells and biological factors, 3D printing provides a revolutionary solution for reconstructing complete osteochondral units. However, achieving effective cartilage repair ultimately requires hydrogel strategies that are specifically tailored to the pathological mechanisms of the injury, which would significantly enhance the outcomes of regeneration.

This review aims to systematically elaborate on recent advances in the use of injectable hydrogels for cartilage repair. It begins by analyzing the structural characteristics, pathological mechanisms, and limitations of existing therapeutic strategies for cartilage damage. The unique advantages of injectable hydrogels as a repair platform are then highlighted, along with a detailed discussion of their application strategies and recent achievements in delivering various therapeutic agents. The synergistic role of 3D printing technology in constructing complex osteochondral scaffolds is also evaluated. Finally, this review summarizes the current challenges in the field and offers perspectives on future directions, with the goal of providing theoretical insights and technical guidance to promote the clinical translation of precision cartilage repair technologies.

2. Osteochondral injury

2.1 Cartilage part

Articular cartilage is a specialized connective tissue that covers the ends of bones in diarthrodial joints. Its intricate structure and composition are engineered to provide mechanical support and exceptional lubrication. The tissue is primarily composed of water, collagen, proteoglycans, and a sparse population of chondrocytes. Water constitutes 65% to 80% of the wet weight, conferring high hydration and elasticity, which enables stress dissipation through fluid movement during compression [1]. Type II collagen forms a 3D network, accounting for 25% to 35% of the dry weight. In the superficial zone, collagen fibers align parallel to the surface to resist shear forces, whereas in deeper zones, they adopt an interwoven, wavy configuration that enhances compressive strength [18]. Proteoglycans, such as aggrecan, which contains chondroitin sulfate and binds to hyaluronic acid, represent 10% to 15% of the dry weight. These molecules trap water to create a molecular sieve, regulate ion balance, and act as biological lubricants. The synergistic interaction between glycosaminoglycan chains and the collagen network confers the distinctive viscoelastic properties of cartilage.

Chondrocytes represent the functional core of cartilage but occupy only 1% to 5% of the tissue volume, residing in cavities known as lacunae. Superficial zone chondrocytes are flattened and metabolically active, synthesizing and renewing the ECM. In the deep zone, chondrocytes form isogenous groups and maintain matrix homeostasis by secreting type II collagen and proteoglycans. Structurally, articular cartilage is organized into four distinct layers [19, 20]. The superficial zone, comprising 10% to 20% of the thickness, contains densely aligned collagen fibers and a low cell density, providing a smooth, low-friction surface. The transitional zone features obliquely oriented collagen fibers and more rounded chondrocytes. The radial zone is characterized by collagen bundles arranged perpendicular to the joint surface, with chondrocytes aligned in columns. This structure culminates at the tidemark, a basophilic line that demarcates the transition to the calcified zone. The calcified zone anchors the cartilage to the underlying subchondral bone through a mineralized interface, thereby facilitating the transfer of mechanical loads **(Figure 1A)** [21]. This layered architecture and gradient composition provide articular cartilage with its remarkable biomechanical properties. A smooth collagen network and synovial fluid containing hyaluronic acid and lubricin maintain an extremely low coefficient of friction, which can be as low as 0.002 during movement. The deep calcified layer transmits mechanical loads to the subchondral bone [22]. Furthermore, the water-retaining capacity of proteoglycans allows the

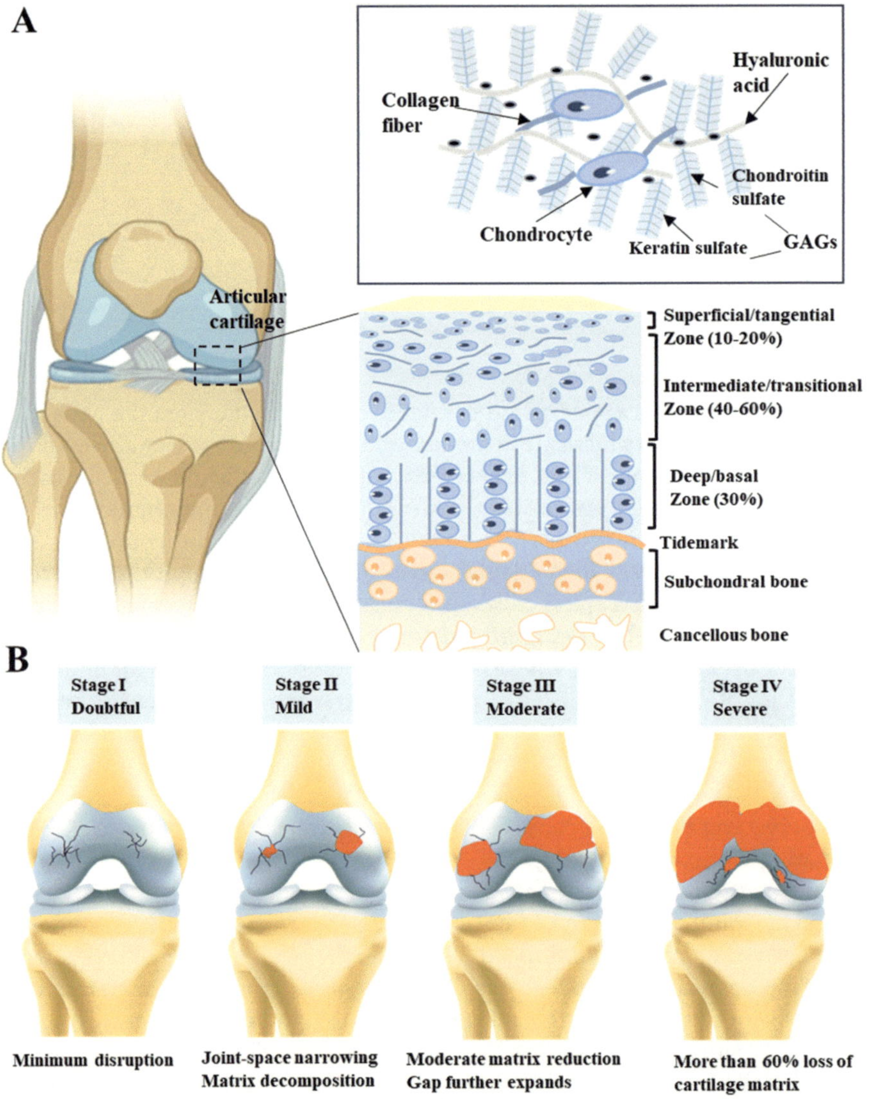

Figure 1.
Structure of articular cartilage and classification of cartilage damage levels.

matrix to harden under compression by expelling fluid, thereby dispersing stress, and to recover its elasticity upon unloading as water is reabsorbed. It is critical to note that articular cartilage is both avascular and aneural. Nutrient supply occurs primarily via diffusion from the synovial fluid, supplemented by permeation from the subchondral bone, while metabolic waste is removed through cyclic loading and absorption [23, 24]. This passive diffusion-limited nutrition mechanism severely constrains the tissue's innate repair capacity. While superficial injuries may see limited chondrocyte-mediated repair, deeper defects typically progress to fibrocartilage formation or calcification.

2.2 Bone part

The subchondral bone serves as a critical structural and metabolic foundation for the overlying articular cartilage. This complex tissue exhibits a distinct hierarchical organization, comprising a dense cortical bone plate at the surface and a porous trabecular bone network beneath it. The cortical bone, with its highly organized lamellar structure of collagen fibers, provides a solid mechanical anchor for the cartilage and effectively resists excessive deformation under load [25]. Underlying this, the trabecular bone forms an optimized mechanical system through its unique lattice of trabeculae. This network demonstrates a high porosity ranging from 50% to 90%, and the spatial orientation of the trabeculae aligns closely with the joint's primary loading direction [26]. This refined architectural design allows the trabecular bone to absorb mechanical energy through elastic deformation while achieving high load-bearing efficiency with minimal weight [27].

A delicate cellular balance maintains the subchondral bone at the cellular level. Osteoblasts function as the primary bone-forming cells, continuously synthesizing key matrix components, including type I collagen and osteocalcin. In contrast, multinucleated osteoclasts are responsible for bone resorption, which they accomplish by secreting acids and proteolytic enzymes to degrade the bone matrix. Osteocytes, embedded within the mineralized matrix in the lacunar-canalicular system, act as the principal mechanosensors. Accounting for over 90% of all bone cells, osteocytes precisely regulate the dynamic balance between bone formation and resorption through an extensive communication network, thereby maintaining skeletal homeostasis [28].

From a materials science perspective, the mechanical properties of the bone matrix originate from its unique composite nature. The organic phase, predominantly type I collagen, constituting 20% to 30% of the dry weight, forms a 3D network that provides toughness and resistance to fracture. The inorganic phase, composed mainly of hydroxyapatite crystals, which account for 60% to 70% of the dry weight, impregnates this collagen network to impart stiffness and compressive strength [29]. This natural composite design enables the subchondral bone to meet the dual mechanical demands of high compressive strength and fracture toughness, making it ideally suited for the long-term weight-bearing function of joints.

Furthermore, the subchondral bone and the articular cartilage constitute a functionally integrated osteochondral unit. Mechanically, the subchondral bone dissipates stress through its controlled deformability, thereby preventing localized stress concentrations in the cartilage. Metabolically, its abundant vascular network provides essential nutrient support for the avascular cartilage via diffusion, helping to sustain chondrocyte viability. This profound structural and functional integration means that pathological changes in the subchondral bone, such as sclerosis or cyst formation, can directly disrupt the cartilaginous microenvironment and accelerate its degeneration. This tight functional coupling provides a critical histological basis for understanding the pathogenesis of osteoarthritis [29, 30].

2.3 Osteochondral defects

Injuries to the articular cartilage and subchondral bone are common orthopedic pathologies characterized by diverse types and complex mechanisms arising from the interplay of mechanical stress, biochemical imbalance, and a pathological microenvironment. Traumatic injury represents the most direct cause, where intense impact or falls during sports can induce superficial cartilage microcracks or subchondral bone bruising. These acute injuries often present with joint swelling and limited mobility and may progress to chronic degeneration without prompt intervention. Degenerative injury is strongly associated with aging and is characterized by reduced chondrocyte metabolic activity, decreased synthesis of type II collagen and proteoglycans, and elevated activity of matrix-degrading enzymes such as MMP-13. These changes lead to cartilage fibrillation and subchondral bone sclerosis, resulting in osteophyte formation and exacerbated joint pain [31]. Inflammatory injury, prevalent in conditions like rheumatoid arthritis, involves cytokines such as IL-1β and TNF-α activating chondrocyte apoptosis pathways. This process accelerates matrix degradation while inhibiting intrinsic repair mechanisms, ultimately causing full-thickness cartilage loss and subchondral bone erosion.

The pathological manifestations of cartilage injury can be categorized based on depth and severity. Superficial injuries involve surface softening or fibrillation, where chondrocytes retain partial metabolic function but possess limited regenerative capacity. Deep injuries extend into the radial and calcified zones, featuring collagen network disruption and abnormal calcification that can provoke stress fractures in the underlying bone. In full-thickness injuries, the complete loss of cartilage exposes the subchondral bone to synovial fluid, often leading to the formation of loose bodies that accelerate osteoarthritis progression. A distinct pathological entity, osteochondritis dissecans, frequently occurs in adolescent knees and stems from localized subchondral ischemia or repetitive microtrauma. This condition causes subchondral bone necrosis, with the affected bone fragment potentially detaching and causing mechanical symptoms like joint locking and chronic pain. For evaluation, magnetic resonance imaging clearly visualizes full-thickness cartilage defects and associated bone marrow edema, while computed tomography provides precise assessment of bony structural integrity, thereby supporting graded treatment planning [32]. The International Cartilage Repair Society grading system offers a quantitative clinical standard for assessing injury severity. This system ranges from grade I, indicating surface softening, to grade IV, representing full-thickness defects with exposed bone, where higher grades correlate with more advanced osteoarthritis and greater challenges for effective repair (**Figure 1B**) [33].

The clinical outcomes of cartilage injury are intrinsically linked to the selected repair strategy. Mild injuries may be managed conservatively through load reduction, physical therapy, or intra-articular injections, such as platelet-rich plasma

(PRP), to modulate the local biological environment. Moderate to severe injuries typically necessitate surgical interventions, including microfracture, autologous chondrocyte implantation, or the transplantation of engineered biomaterial scaffolds. Recent advances in stem cell-based tissue engineering have focused on replicating the native cartilage's mechanical gradient structure, which has significantly improved the functional integration of regenerated tissue and offers promising directions for treating complex defects. Nevertheless, significant challenges persist, including the prevention of abnormal subchondral vascularization and achieving a biomechanical match between the repaired and native tissue. The development of multimodal therapeutic strategies that address these multifaceted issues remains essential for optimizing long-term clinical outcomes.

3. Pathological mechanisms of cartilage injury and therapeutic limitations

3.1 Pathological mechanisms of cartilage injury

The pathological mechanisms of cartilage injury involve complex cellular and molecular interactions among cartilage, the synovium, and the subchondral bone. In the initial stage, mechanical stress or inflammatory stimuli trigger chondrocytes to release matrix-degrading enzymes such as matrix metalloproteinases and aggrecanases. This enzymatic activity progressively breaks down collagen fibers and proteoglycans, leading to structural disruption and functional impairment of the ECM. Concurrently, oxidative stress induces mitochondrial dysfunction in chondrocytes, resulting in the accumulation of reactive oxygen species. This oxidative damage activates caspase-mediated apoptosis and stimulates key signaling pathways, including NF-κB and MAPK, which promote the further release of pro-inflammatory cytokines like IL-1β and TNF-α [34]. These inflammatory factors then amplify the immune response by activating macrophages and synovial fibroblasts within the synovial cavity. This activation leads to the overexpression of destructive enzymes such as MMP-13 and ADAMTS-5, which accelerate the breakdown of the core cartilage components. Furthermore, matrix degradation products, including hyaluronic acid fragments and collagen peptides, function as damage-associated molecular patterns. These fragments persistently activate synovial immune cells through pattern recognition receptors such as TLR4, thereby establishing a self-perpetuating positive feedback loop that sustains chronic inflammation [35].

Simultaneously, the subchondral bone undergoes compensatory remodeling characterized by increased osteoclast activity and bone resorption, alongside attempted repair by osteoblasts. During this remodeling process, factors such as transforming growth factor-β and insulin-like growth factor are released. These factors not only drive bone changes but also permeate into the overlying cartilage layer, where they exert a direct influence on chondrocyte metabolism [36]. The pathology is further propagated as macrophages and synoviocytes in the synovial lining recognize cartilage debris, activating the NF-κB pathway and secreting more pro-inflammatory cytokines, which diffuse back into the cartilage.

As the disease advances, blood vessels from the subchondral bone penetrate the calcified cartilage layer, forming an invasive vascular pannus [37, 38]. This invasion disrupts the critical integrity of the osteochondral junction and introduces additional inflammatory cells and pro-angiogenic factors into the cartilage environment. This process establishes a vicious cycle intertwining inflammation, matrix degradation, and abnormal vascularization. Ultimately, this self-reinforcing tripartite pathological network among cartilage, synovium, and subchondral bone results in irreversible cartilage loss, chronic synovitis, and aberrant bone remodeling, collectively driving the progression of osteoarthritis (**Figure 2**).

3.2 Limitations of current pharmacological and surgical treatments

Current treatment strategies for osteoarthritis are designed to alleviate symptoms and slow disease progression, but they generally fail to reverse existing cartilage damage. These approaches exhibit significant limitations in their therapeutic precision, long-term safety, and broad clinical applicability. Pharmacological interventions primarily involve non-steroidal anti-inflammatory drugs and intra-articular injections [39, 40]. While these agents can suppress inflammation in the short term, for example, by reducing prostaglandin E2 production through COX-2 inhibition, their inability to penetrate the dense cartilage matrix prevents them from reaching the core of the injury. Furthermore, prolonged use of these drugs may paradoxically accelerate cartilage calcification. Biological agents, such as IL-1 receptor antagonists or anti-TNF-α monoclonal antibodies, offer more targeted modulation of inflammatory pathways. However, they carry a risk of

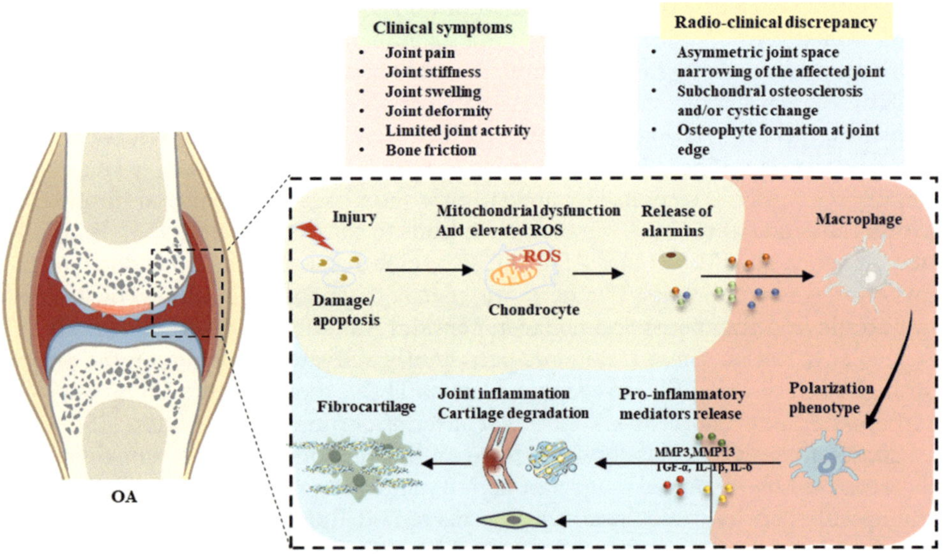

Figure 2.
Pathological progression and key mechanisms of osteoarthritis. Common clinical symptoms include joint pain, stiffness, swelling, deformity, limited range of motion, and bone friction. The molecular cascade is initiated by joint injury, leading to mitochondrial dysfunction and elevated ROS. This triggers the release of alarmins, which activate macrophages and influence their polarization phenotype. Activated macrophages secrete pro-inflammatory mediators such as TGF-β, IL-1β, and IL-6, which, in turn, promote the expression of matrix-degrading enzymes, including MMP-3 and MMP-13. These processes collectively drive chondrocyte damage and apoptosis, cartilage degradation, joint inflammation, and ultimately result in the formation of mechanically inferior fibrocartilage.

immunogenicity and demonstrate limited efficacy in restoring an ECM that has already been degraded. Gene therapy approaches that utilize viral vectors to deliver anabolic genes like SOX9 or BMP aim to enhance matrix synthesis. Their clinical translation remains challenging due to low delivery efficiency, uncontrollable transgene expression, and persistent safety concerns [41]. A summary of typical drugs and small molecules used in cartilage injury treatment is provided in **Table 1.**

In the realm of surgical interventions, microfracture techniques promote the formation of fibrocartilage by recruiting bone marrow stem cells. The resulting repair tissue, however, lacks the biomechanically robust architecture of native hyaline cartilage and exhibits a high failure rate of up to 60% within two to three years [42]. Autologous osteochondral transplantation can provide genuine hyaline cartilage, but it is associated with donor site morbidity in 20% to 30% of cases. Furthermore, the repair of large defects often requires multiple complex procedures. Allograft transplantation carries inherent risks of immune rejection due to challenges in HLA matching and potential disease transmission, with five-year graft survival rates reported below 70%. Autologous chondrocyte implantation necessitates two separate surgical procedures and entails high costs, while the survival rate of the implanted chondrocytes is only 30% to 50%. These cells also frequently lose their stable phenotype over the long term. Emerging technologies, such as 3D bioprinted scaffolds, can mimic the mechanical gradients of the native ECM. However, the inadequate vascularization of these constructs often leads to central cell necrosis, and the engineered tissues frequently fail to integrate and adapt under dynamic mechanical loading in the joint [43, 44]. These collective shortcomings underscore a fundamental inadequacy of existing therapies to achieve targeted, durable, and functionally integrated cartilage repair. This persistent challenge highlights the urgent need for the development of multi-modal therapeutic strategies that effectively combine biological activity with biomechanical compatibility.

4. Advantages of injectable hydrogel for cartilage repair

Injectable hydrogels present multifunctional platforms for cartilage repair, offering innovative solutions based on their minimally invasive delivery, capacity for controlled therapeutic release, mechanical adaptability, and inherent lubricating properties (**Figure 3**). Their primary clinical advantage is the ability for precise, minimally invasive injection that conforms to and fills irregularly shaped defects within the joint cavity [55, 56]. This in situ gelation process eliminates the need for open surgical procedures. The resulting 3D network achieves dynamic mechanical compatibility with the native tissue, as the hydrogel stiffness can be finely tuned from 0.1 to 100 kPa to match the elastic modulus of natural cartilage [57], which is approximately 1 to 10 MPa [55]. Advanced designs that incorporate a calcified layer interface further facilitate physiological stress transmission to the underlying subchondral bone, thereby preventing secondary damage from mechanical mismatch.

The porous 3D structure formed after cross-linking effectively recapitulates the physical architecture of the native cartilage ECM. This biomimetic scaffold provides essential physical support for cell colonization and proliferation while simultaneously enhancing the diffusion of nutrients and metabolic waste, thereby reconstructing a conducive microenvironment for functional cartilage regeneration [58–60]. Regarding therapeutic delivery, the chemically cross-linked network of

Typical molecule/drug	Indications	Mechanism	Outcomes	Disadvantages	Reference
Methotrexate	Knee osteoarthritis (KOA)	Exact anti-inflammatory mechanism not fully elucidated; may involve suppression of immune cell activity.	Oral administration significantly alleviates pain and stiffness and improves function in KOA patients.	As an immunosuppressant, long-term safety (e.g., hepatotoxicity) requires further evaluation in OA populations.	[45]
Lorecivivint	OA	Small-molecule CLK/DYRK1A inhibitor that modulates Wnt and inflammatory pathways.	Currently in phase III clinical trials for KOA.	Still under clinical investigation; long-term efficacy and safety require final confirmation.	[46]
Fasinumab	OA	Inhibitor of nerve growth factor (NGF).	Provides a novel mechanism of action for OA pain management.	Associated with the risk of adjudicated arthropathy, primarily rapidly progressive OA.	[47]
Curcumin	OA/RA	Multi-modal actions: inhibits NF-κB and COX-2 pathways (anti-inflammatory); scavenges ROS (antioxidant); inhibits MMPs and promotes type II collagen and aggrecan synthesis (chondroprotective).	Offers multiple benefits as a natural polyphenol: anti-inflammatory, antioxidant, anti-catabolic, and chondroprotective.	Poor water solubility, low oral bioavailability, rapid metabolism; highly dependent on advanced delivery systems.	[48]
Bergapten	RA	Natural small molecule that directly binds STAT3 and inhibits its phosphorylation, thereby blocking inflammatory crosstalk between macrophages and synovial fibroblasts.	Provides a novel STAT3-targeting candidate for RA treatment with low cytotoxicity to normal cells.	Requires structural optimization to improve targeting and bioavailability.	[49]
Coelonin (COE)	RA	Natural dihydrophenanthrene compound acting as a natural Src kinase inhibitor; simultaneously inhibiting osteoclastogenesis and the inflammatory activation of synovial fibroblasts.	Potential as a dual-function anti-RA candidate, inhibiting both bone resorption and synovitis.	Currently in the preclinical stage; further clinical studies are needed to verify efficacy and safety in humans.	[50]
UBX0,101	OA	Targets and clears senescent cells to inhibit cartilage destruction.	Represents a potential senolytic strategy for OA treatment.	Still in clinical trials; final efficacy and safety require confirmation.	[51]
BNTA	OA	Activates SOD3 pathway to reduce superoxide anions, inhibiting ECM degradation and the inflammatory response.	Preclinical trials show repair efficiency superior to microfracture.	Complex preparation process; high cost for large-scale production.	[52]
10-had	Osteoarthritis cartilage degeneration	Targets the Asp_Arg_Hydrox domain of ASPH, inhibiting the ERK/p53/p21 pathway to delay chondrocyte senescence.	Effective via both oral and intra-articular administration, but long-term safety data is insufficient.	Systemic toxicity beyond local closed delivery requires verification.	[53]
Melittin	Osteoarthritis inflammation regulation	Delivered via zirconium metal-organic framework (Zr-MOF); inhibits COX-2 and NF-κB pathways to reduce PGE2 release.	Local injection extends pain relief duration up to 12 weeks.	Initial release phase may cause transient cytotoxicity.	[54]

Table 1.
Mechanisms and defects of small molecules or drugs used for cartilage injury repair.

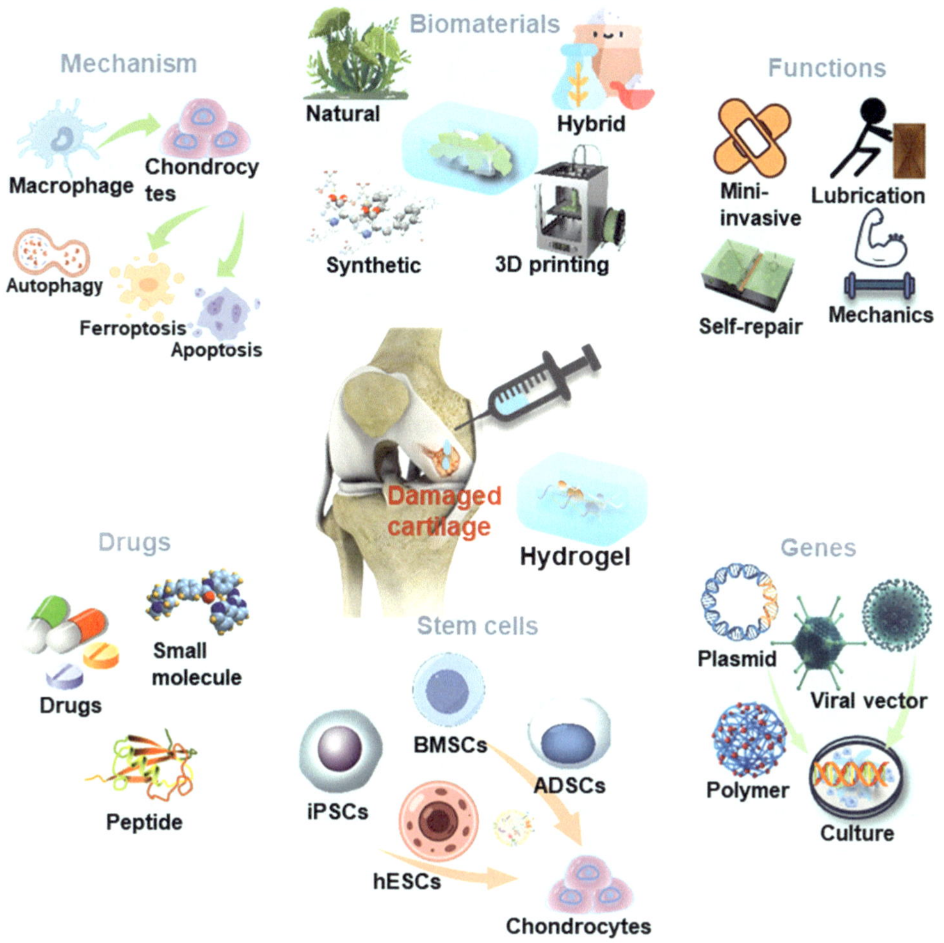

Figure 3.
Schematic diagram of injectable hydrogel used as a therapeutic agent repository for articular cartilage repair. Hydrogels offer several distinct advantages as scaffolds for cartilage repair. Their injectable nature undergoing a sol-gel transition enables minimally invasive precise delivery of therapeutic agents to injury sites. This approach prevents swelling damage caused by bone-on-bone friction and reduces patient discomfort. These materials possess tunable mechanical properties and exceptional shape adaptability making them ideal for filling both large and small irregular cartilage defects while serving as load-bearing biomimetic structures. Functioning as reservoirs for therapeutic agents' hydrogels can be engineered to respond intelligently to the local microenvironment enabling precise spatiotemporally controlled loading and release of bioactive molecules for targeted treatment. Stem cells released from hydrogels promote cartilage regeneration by differentiating into chondrocytes or secreting exosomes that enhance chondrocyte proliferation. Genes released from hydrogels silence pathogenic factors or increase the expression of repair genes within infected chondrocytes to facilitate cartilage regeneration.

hydrogels allows for the efficient encapsulation of diverse bioactive cargo. Thermosensitive hydrogels, such as those utilizing poly(N-isopropylacrylamide), undergo a sol-gel transition at physiological temperature to form a physical barrier that can extend drug half-life beyond two weeks [61]. Alternatively, the incorporation of pH-sensitive chemical bonds enables an inflammation-responsive drug release profile, maintaining a therapeutically effective anti-inflammatory concentration precisely at the injury site [62].

Specialized conductive hydrogels, for instance, systems modified with polypyrrole, can synergize with applied electrical stimulation to promote the chondrogenic differentiation of stem cells, thereby guiding the formation of metabolically active neocartilage [63]. Lubrication is another critical feature enhanced through biomimetic design; hydrogels modified with sulfonated hyaluronic acid can form a robust surface hydration layer, achieving an ultra-low coefficient of friction of approximately 0.003, which is comparable to the performance of natural articular cartilage. Furthermore, hydrogels with dynamic cross-links possess self-healing capabilities, allowing them to autonomously repair micro-damage incurred during joint movement and thus maintain a smooth articulating surface. Composite systems, such as hyaluronic acid-chitosan lubricating films, function as a physical barrier against invasive fibrous tissue while also adsorbing endogenous lubricin from the synovial fluid to establish a mechanism for sustained lubrication [64]. This multi-mechanism synergy enables hydrogels to maintain their functional integrity under repetitive joint loading, thereby creating a stable and protective biomechanical environment conducive to long-term regeneration.

Finally, the degradation profile of an engineered hydrogel can be precisely programmed through rational material design. This ensures that the scaffold resorption rate is synchronized with the pace of new tissue formation, providing temporary mechanical support before undergoing safe metabolic clearance without necessitating secondary surgical removal. This controlled life cycle completes a holistic therapeutic strategy for cartilage repair.

5. Precision design of hydrogels for cartilage damage in joints

The pathological microenvironment after cartilage injury demonstrates a range of characteristic biochemical and biomechanical alterations. These changes offer clear targets for developing precision-responsive smart hydrogels. First, in response to the typical acidic pH, elevated matrix metalloproteinases (MMPs), and reactive oxygen species (ROS) found in injured areas, stimulus-responsive hydrogels can be engineered [65, 66]. For example, employing pH-sensitive chemical bonds, such as hydrazone bonds or Schiff base bonds, enables specific degradation and controlled release of anti-inflammatory drugs within the acidic inflammatory environment [67]. Incorporating MMP-specific substrate peptide sequences into the hydrogel cross-linked network allows enzyme-triggered degradation and drug delivery, thereby accurately targeting zones with active extracellular matrix breakdown [68]. Furthermore, integrating ROS-scavenging groups, like boronate esters or thioethers, or utilizing nanozymes helps neutralize excess ROS [69]. This action mitigates oxidative stress damage to chondrocytes and establishes a protective reductive setting for tissue regeneration.

Second, the osteochondral unit, consisting of cartilage and subchondral bone, possesses distinct mechanical properties and biological functions. This demands a layered biomimetic approach in hydrogel design [70]. Using 3D printing or sequential injection methods, bilayer or gradient hydrogel scaffolds can be constructed. The segment oriented toward the cartilage layer is generally designed as a soft hydrogel with a low compressive modulus, approximately 0.1 to 1 MPa [71]. It is enriched with hyaluronic acid or type II collagen to imitate natural cartilage mechanics and encourage cartilage formation. The portion facing the subchondral bone requires the addition of hydroxyapatite nanoparticles or hydrogels with higher cross-linking densities to achieve greater stiffness on the gigapascal scale [72]. This supplies necessary mechanical cues for bone formation and supports bone tissue integration. The gradual transition in mechanical and biochemical properties from the cartilage layer to the bone layer effectively mimics the native osteochondral interface. It fosters the development of functionally integrated tissue and prevents mechanical incompatibility and repair failure at the junction [73].

In summary, by precisely aligning the physicochemical characteristics of hydrogels, including degradation kinetics, mechanical strength, and bioactivity, with the specific signals in the cartilage injury microenvironment (such as pH levels, matrix metalloproteinases, and reactive oxygen species), and the anatomical hierarchy between cartilage and bone layers, researchers can transition from a universal one-size-fits-all treatment to a tailored, on-demand precision strategy for cartilage regeneration.

6. Application of therapeutic molecule-loaded hydrogel in cartilage repair

6.1 Drug-loaded hydrogel for cartilage repair

Drug-loaded hydrogels represent an advanced strategy for cartilage repair in osteoarthritis, capitalizing on their 3D network structure, excellent biocompatibility, and controllable drug release profiles. These biomaterial systems serve a dual function by acting as biomimetic scaffolds that replicate the native cartilage ECM to provide mechanical support for new tissue growth, while also enabling the targeted delivery of therapeutic agents to modulate the pathological microenvironment and enhance chondrocyte activity. This combined approach helps disrupt the self-perpetuating cycle of cartilage degradation and chronic inflammation.

Recent research has demonstrated substantial progress in utilizing drug-loaded hydrogels for cartilage regeneration. Scientists have engineered systems that incorporate small molecule drugs, targeted pathway inhibitors, and clinically relevant agents to achieve a coordinated therapeutic outcome combining microenvironment regulation with structural tissue repair. For instance, Chen et al. [74] developed an inflammation-responsive hydrogel loaded with bevacizumab for anti-inflammatory and anti-angiogenic effects in cartilage repair. This system employs a dynamic hydrogen-bonded network stabilized by phenolic hydroxyl groups to enable sustained release of the anti-angiogenic drug at the defect site, effectively inhibiting VEGF signaling and preventing pathological vascular invasion. Concurrently, released magnesium ions mitigated inflammation through suppression of the NF-κB pathway while promoting chondrogenic differentiation of resident bone marrow mesenchymal stem cells. In a rabbit model of full-thickness cartilage defects, this treatment generated regenerated tissue with collagen organization closely resembling that of native hyaline cartilage.

Addressing chondrocyte ferroptosis, Yang et al. created a folic acid-modified polydopamine nanoparticle hydrogel delivering leonurine [75]. This targeted system significantly reduced IL-1β-induced ferritinophagy by specifically modulating M1 macrophage polarization. The gelatin-polyethylene glycol diacrylate composite formed a durable drug reservoir within the joint space, extending the therapeutic duration to more than three times that of conventional delivery methods while restoring the elastic modulus of repaired cartilage to near-normal levels.

Huang et al. designed a gel responsive to reactive oxygen species that encapsulates cartilage growth factor (KGN) for cartilage repair [76]. This system underwent ultrasound-triggered in situ gelation with subsequent sustained KGN release, which activated the Smad5/mTOR pathway to drive chondrogenic differentiation of mesenchymal stem cells. In rabbit models, this treatment achieved 78% depth repair in full-thickness defects, with glycosaminoglycan content in the regenerated tissue reaching 92% of normal cartilage values. Furthermore, a calcium/strontium co-crosslinked hydrogel incorporating the Wnt pathway inhibitor LGK-974 effectively suppressed β-catenin signaling and reduced the proportion of pro-inflammatory M1 macrophages to 15% [77]. The incorporated strontium ions simultaneously stimulated chondrocyte anabolic activity, resulting in 68.5% type II collagen-positive areas in a porcine cartilage defect model, presenting a promising approach for osteoarthritis intervention.

These advancements highlight the multidimensional benefits of drug-loaded hydrogel platforms, where small molecule drugs provide prolonged therapeutic action through sustained release kinetics, targeted inhibitors precisely block pathological signaling cascades, and clinically approved compounds facilitate translational potential. From a technical perspective, dynamic crosslinked networks enable stimulus-responsive drug release, biomimetic mechanical gradients replicate zonal cartilage characteristics, and surface modifications enhance biocompatibility and integration. Future development will likely focus on sophisticated multi-drug co-delivery systems capable of spatiotemporally controlled sequential release, such as initially releasing anti-inflammatory agents followed by pro-regenerative factors, or designing intelligent responsive hydrogels that synchronize drug release profiles with the distinct phases of the tissue regeneration process.

6.2 Stem cell-loaded hydrogels for cartilage repair

Stem cell-loaded hydrogels represent a rapidly advancing frontier in cartilage tissue engineering, combining the regenerative potential of various stem cell sources with the supportive and biomimetic properties of hydrogel scaffolds. These integrated systems have demonstrated significant progress in achieving functional cartilage regeneration. Researchers have successfully incorporated multiple cell types, including bone marrow mesenchymal stem cells (BMSCs), adipose-derived mesenchymal stem cells (ADSCs), induced pluripotent stem cells (iPSCs), and human embryonic stem cells (hESCs), into hydrogel matrices to enhance repair outcomes. For instance, Zhong et al. [78] developed an injectable hydrogel derived from decellularized porcine meniscus ECM and loaded it with BMSCs for implantation in a rabbit model of full-thickness white-zone meniscus injury. The thermosensitive hydrogel, combined with the immunomodulatory properties of the ECM, synergistically enhanced chondrogenic differentiation of the stem cells, resulting in regenerated tissue with glycosaminoglycan content comparable to native cartilage. In another approach, Wang et al. created a methacrylated silk fibroin hydrogel featuring surface microgrooves to guide the oriented growth of BMSCs [79]. When combined with a decellularized cartilage matrix scaffold in a rabbit tracheal defect model, this construct supported the regeneration of neocartilage with a native-like elastic modulus after 12 weeks, without calcification or fibrosis.

Adipose-derived mesenchymal stem cells have gained prominence due to their accessibility and abundance. Yang et al. [80] designed a chitosan hydrogel composite loaded with ADSCs, in which an optimized crosslinked network enabled the sustained release of the cells. This system significantly enhanced type II collagen expression while reducing type I collagen deposition, indicating successful inhibition of fibrotic tissue formation. In a more sophisticated strategy, Chen et al. incorporated human urine-derived stem cells overexpressing basic fibroblast growth factor into a decellularized small intestinal submucosa hydrogel for tendon-bone interface repair [81]. This construct modulated local inflammation by promoting M2 macrophage polarization, resulting in extensive type II collagen deposition covering 68.5% of the defect area and mechanical properties restored to near-normal levels.

hESCs offer exceptional differentiation potential but face clinical limitations due to ethical concerns and tumorigenic risks. As a promising alternative, induced pluripotent stem cells have emerged as a viable cell source. Toh et al. [82] demonstrated that hESC-derived chondrocytes encapsulated in hyaluronic acid (HA) hydrogels formed ECM-rich cartilage structures. When implanted into critical-sized osteochondral defects in rats, these constructs closely mimicked native osteochondral tissue after 12 weeks, while control defects treated with hydrogel alone showed poor repair. Importantly, no tumor formation was observed, supporting the safety profile of this approach. Similarly, porcine model studies confirmed that transplanted iPSCs could regenerate cartilage at osteochondral defect sites without tumorigenesis [83]. A 3D microchamber hydrogel transition platform has further been developed to guide iPSC differentiation into chondrocytes, producing grafts rich in type II collagen. Notably, the repair quality achieved with iPSC-derived chondrocytes surpassed that mediated by BMSCs, potentially due to epigenetic modifications such as increased methylation at CpG sites in the COL10A1 promoter that suppress hypertrophic differentiation and promote stable cartilage phenotypes. Despite these advances, low survival rates of transplanted iPSCs remain a significant challenge.

In summary, stem cell-loaded hydrogels have demonstrated promising cartilage repair outcomes in animal models. However, clinical translation requires overcoming several obstacles, including mismatched degradation and regeneration kinetics, unstable chondrogenic differentiation of stem cells, and the establishment of standardized manufacturing protocols. Future research should focus on advanced material modifications, precise control of release kinetics, and optimized preclinical evaluation systems to accelerate the clinical application of these regenerative strategies.

6.3 Gene-loaded hydrogels for cartilage repair

Gene-loaded hydrogels represent an advanced therapeutic strategy for multi-dimensional intervention in cartilage degeneration through the incorporation of novel regulatory molecules, including miRNA, siRNA, and circRNA. These systems enable precise genetic modulation within the challenging joint environment. Li et al. [84] developed a streamlined zinc oxide hydrogel delivering miR-17-5p through a dynamically crosslinked network. This design reduced injection resistance by 40% due to its optimized hydrodynamic properties, while the responsively released miRNA effectively inhibited MMP-13 expression and activated the chondrogenic

SOX9 pathway. In a rat model, this treatment restored the elastic modulus of regenerated cartilage to 85% of normal tissue values. To address cellular senescence, a self-assembling peptide hydrogel delivering miR-29b-5p was engineered to recruit synovial mesenchymal stem cells through incorporation of the homing peptide SKPPGTSS [85]. The sustained miRNA release suppressed P16INK4a expression via the TET1/PPAR γ pathway, achieving 78% depth repair of full-thickness cartilage defects in an osteoarthritic rat model. Similarly, Long et al. created a nanocomposite hydrogel loaded with miR-455-3p that utilized a biomimetic gradient pore structure to enable sustained release [86]. This system improved miRNA delivery efficiency threefold, significantly enhanced type II collagen synthesis, and inhibited ADAMTS5 activity, resulting in glycosaminoglycan content recovery reaching 92% of normal levels in a porcine cartilage defect model.

In the domain of siRNA delivery, an inflammation-regulated self-assembling hydrogel microsphere system was developed to deliver siRNA targeting IL-1β [87]. By incorporating pH-responsive calcium phosphate co-precipitation, this system achieved 80% gene silencing efficiency in the weakly acidic microenvironment of cartilage while effectively anchoring to the cartilage matrix to prevent therapeutic diffusion. In a rabbit model, it reduced MMP-13 expression by 62%. Zhang and collaborators constructed a metal-organic framework co-loaded with HIF-2α siRNA and curcumin [87]. The released siRNA effectively cleaved HIF-2α mRNA, achieving potent gene silencing, while curcumin concurrently suppressed inflammation and cartilage degradation, demonstrating a valuable synergistic therapeutic effect.

The repair mechanisms of gene-loaded hydrogels primarily operate through three key pathways. The first pathway involves inflammation suppression, exemplified by microRNA-140-loaded hydrogels that significantly reduce intra-articular levels of pro-inflammatory factors, including IL-6, IL-1β, and TNF-α. The second pathway promotes chondrogenic differentiation, as demonstrated by systems containing small molecule inducers like TD-198946, which effectively drive mesenchymal stem cells toward chondrogenic commitment while inhibiting fibrotic changes. The third mechanism focuses on maintaining ECM homeostasis by regulating the expression of collagen-degrading enzymes, such as MMP-13 and ADAMTS-5, thereby reducing matrix breakdown.

Despite these significant advances, several challenges remain unresolved. Non-viral delivery systems frequently suffer from limited transfection efficiency, while viral vectors present potential safety concerns, including immunogenicity and insertional mutagenesis. Additionally, mismatched kinetics between hydrogel degradation and tissue regeneration rates often compromise clinical outcomes. Successful translation to clinical applications will require overcoming several hurdles, including validation in large animal models, comprehensive long-term safety assessments, and the development of standardized manufacturing protocols. Future research should prioritize optimizing vector design, developing smart, responsive hydrogels with precise release kinetics, and conducting multi-center clinical trials to advance this promising therapeutic approach toward clinical reality.

7. Application of 3D printing in cartilage repair

3D printing technology has emerged as a promising solution to overcome the limitations of existing cartilage repair strategies and achieve biomimetic regeneration that restores both structural and functional properties. This advanced manufacturing approach enables the precise spatial arrangement of biomaterials, living cells, and bioactive factors, thereby creating new opportunities for constructing personalized bioactive scaffolds that replicate the native osteochondral gradient structure [88].

3D bioprinting technologies offer distinct advantages for cartilage repair applications based on their unique operational principles. Extrusion-based bioprinting represents the most widely utilized technique, employing pneumatic or mechanical systems to deposit bioinks in continuous filaments through layer-by-layer accumulation. This method accommodates a broad spectrum of materials, ranging from high-strength synthetic polymers to soft cell-laden hydrogels, while multi-printhead configurations enable the fabrication of heterogeneous constructs crucial for osteochondral interfaces. For instance, researchers have successfully fabricated mechanically reinforced scaffolds by co-printing gelatin methacryloyl hydrogels containing chondrocytes alongside polycaprolactone structural fibers, demonstrating effective cartilage repair in animal studies [89, 90]. The limitations of this technology include relatively low printing resolution and the potential for reduced cell viability due to shear stress during extrusion. In contrast, vat polymerization bioprinting, such as digital light processing, operates on a projection principle where light selectively solidifies photosensitive resins with high precision. This approach offers significant advantages, including micron-scale resolution and smooth surface finishes that effectively mimic the low-friction characteristics of native articular cartilage, while its layer-projection mechanism substantially accelerates manufacturing speed. Practical applications include digital light processing-printed poly(ethylene glycol) diacrylate and polycaprolactone composite scaffolds for repairing large cartilage defects in goats, with elastic moduli tunable between 0.5 and 5 MPa through material ratio adjustments to match regional mechanical requirements. A primary challenge for this technology involves developing cytocompatible photoinitiators and resins to address the cytotoxicity concerns associated with conventional materials. Additionally, inkjet bioprinting employs precise control mechanisms to generate extremely fine bioink droplets for targeted deposition, offering advantages in printing velocity and cell survival rates. Representative applications include printing microdroplets containing transforming growth factor-β3 into specific regions of hydrogel scaffolds to create localized chondrogenic signal gradients that efficiently direct stem cell differentiation [91]. This method, however, imposes stringent requirements on bioink viscosity and cell concentration, while the resulting constructs typically demonstrate inadequate mechanical strength, rendering it more suitable for precise bioactive factor delivery or combination with other printing technologies.

Leveraging these distinctive technical capabilities, researchers have developed three primary biomimetic strategies aimed at achieving integrated structural and functional regeneration through accurate simulation of native osteochondral tissue. The first strategy employs multi-material gradient printing using distinct bioinks to represent cartilage and bone layers, thereby creating a seamless transition from compliant to stiff regions in the vertical dimension that mimics the natural mechanical progression. The second strategy involves the precise engineering of porous architectures to establish distinct microenvironments appropriate for the biological functions of cartilage and bone layers, while combining rigid frameworks with soft hydrogels through co-printing to ensure mechanical integrity. The final strategy utilizes cell printing techniques to position different functional cells within their respective anatomical regions accurately, employing microcarriers for controlled release of growth factors to spatiotemporally guide zonal tissue regeneration. These complementary approaches collectively advance osteochondral repair toward the goal of functional tissue restoration **(Figure 4)**.

8. Key technical parameters and performance

In injectable hydrogel design, performance hinges on the synergy between response characteristics, mechanical properties, and release kinetics. For instance, thermosensitive hydrogels, such as those based on pNIPAM, typically exhibit gelation times ranging from tens of seconds to several minutes [92]. Their elastic modulus, approximately 1–10 kPa, readily matches that of superficial native cartilage. However, their drug release often relies on passive diffusion, resulting in a relatively short half-life of several days. In contrast, chemically crosslinked or double-network hydrogels can achieve a broader modulus range, from 10 kPa to several megapascals, by adjusting crosslinking density [93]. This enables them to mimic the mechanical gradient from cartilage to subchondral bone. Nevertheless, their gelation time requires precise control to prevent syringe clogging. Smart responsive hydrogels, such as those reacting to ROS, pH, or enzymes, achieve on-demand precise control of drug release [94, 95]. For example, in an inflammatory environment with ROS concentrations around 100 μM, their degradation and drug release rates can increase by three to fivefold [96]. This extends the maintenance of therapeutic doses from several weeks to months, significantly enhancing treatment precision and efficiency.

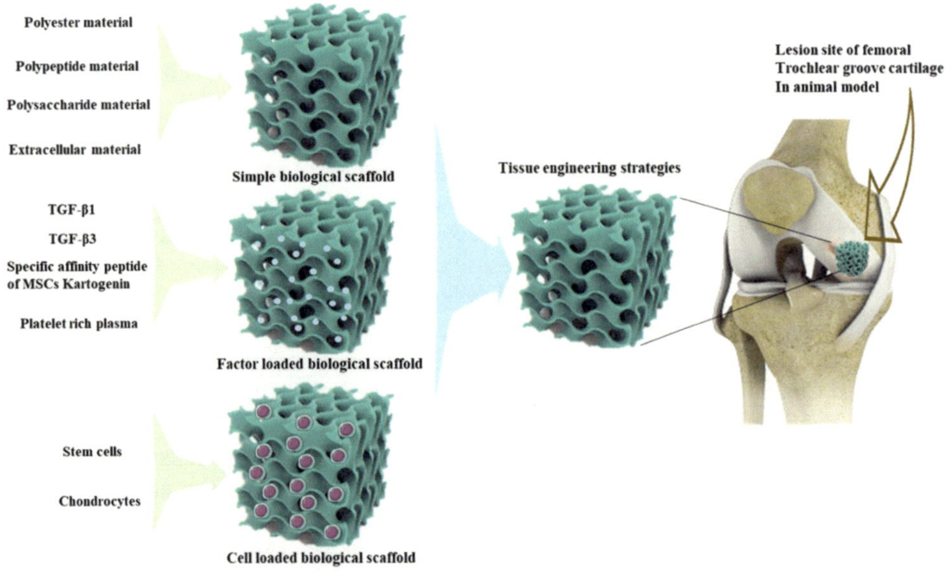

Figure 4.
Schematic diagram of the application of 3D printing in joint cartilage repair.

In selecting 3D printing technologies, different technical pathways present inherent trade-offs among resolution, printing speed, and biocompatibility. Extrusion-based printing can process high-viscosity bioinks and construct mechanically robust scaffolds with compressive moduli reaching hundreds of megapascals [97]. However, its resolution, typically exceeding 100 μm, and cell viability, around 80%, are limited [98]. Vat polymerization printing, such as digital light processing (DLP), can fabricate structurally precise scaffolds with higher resolution, approximately 10–50 μm, and faster speed, with layer curing times in seconds [99]. This makes it highly suitable for mimicking the low-friction characteristics of cartilage surfaces. Yet, the cytotoxicity of commonly used UV photoinitiators and resin monomers remains a critical challenge. Inkjet printing demonstrates excellent cell viability, over 90%, and enables precise distribution of growth factors to create biochemical gradients [100]. However, constrained by low-viscosity inks, the mechanical strength of its formed structures is often inadequate.

In summary, the multiple technologies discussed in this review are not mutually exclusive but rather constitute a toolbox from which researchers can select precision solutions based on specific repair requirements. Future innovation lies in intelligent integration – for example, using stimulus-responsive hydrogels as bioinks for 3D printing. This approach would create living implants that perfectly fit defect shapes while dynamically responding to changes in the internal microenvironment, ultimately advancing cartilage regeneration from static replacement to a dynamic regulation paradigm.

9. Clinical translation challenges for hydrogels

Although hydrogel-based drug, cell, and gene therapies demonstrate promising cartilage repair potential in preclinical animal models such as rabbits, rats, and pigs, their translation toward broad clinical application faces several critical barriers. First, regarding the translation pathway, current research predominantly follows a linear model progressing from small animals for concept validation to large animals for functional and dimensional verification [101]. However, a significant gap exists when transitioning to human trials. For instance, the spontaneous repair capacity and joint mechanical environment of rodents differ substantially from humans [102]. Even large animal models exhibit differences in joint size, cartilage thickness, and life cycle compared to humans, making it difficult to directly extrapolate therapeutic efficacy and dosage parameters from animal studies to human applications.

Second, each therapeutic modality presents unique translational limitations. For drug-loaded hydrogels, the challenges involve ensuring the long-term stability of active molecules within the complex human joint cavity and achieving precisely controlled release kinetics that match the prolonged timeline of human cartilage repair, thereby avoiding initial burst release or premature depletion [103, 104]. The core issues for cell-laden hydrogels include cell source selection, tumorigenicity risks (particularly with iPSC or hESC-derived cells), the cost and quality control of large-scale expansion, and the low survival rate and phenotypic instability of transplanted cells within the inflammatory human environment [105]. Gene-loaded hydrogels face even greater regulatory hurdles and safety concerns, including the immunogenicity of viral or non-viral vectors, potential off-target effects, and insertional mutagenesis, all of which significantly impede their clinical translation pace [105].

Therefore, advancing the clinical translation of hydrogel-based cartilage repair technologies must focus on several key areas. First, standardized manufacturing and quality control are essential to ensure consistent physicochemical properties (such as modulus, porosity, and degradation rate), along with bioactivity across hydrogel batches, meeting good manufacturing practice requirements [106]. Second, conducting more clinically relevant preclinical studies, for example, evaluating efficacy in aged or metabolically impaired animal models, such as those with diabetes, can better simulate the pathological conditions of human osteoarthritis patients. Third, adopting translational medicine principles early, engaging with regulatory agencies to establish evaluation criteria, and designing patient-centered clinical trial endpoints that prioritize pain relief, functional recovery, and quality of life improvements as patient-reported outcomes, alongside imaging and histological assessments [107]. Only through systematic interdisciplinary collaboration to address these translational bottlenecks can hydrogel technology truly advance from laboratory research to surgical practice, becoming a precision medical tool that transforms the clinical management of osteoarthritis.

10. Prospects and conclusions

Although injectable hydrogels and 3D printing technologies show great promise for cartilage repair, current research exhibits significant limitations. These limitations, however, represent the most compelling directions for future innovation. First, contemporary materials possess only rudimentary smart functionality. Most hydrogels face an inherent conflict between mechanical strength and nutrient diffusion efficiency. Furthermore, their degradation rates often fail to synchronize with the prolonged timeline of tissue regeneration. Future investigations should progress beyond optimizing individual properties and instead focus on developing adaptive hydrogel systems [107]. For instance, employing dynamic covalent chemistry, such as Diels-Alder reactions or boronate esters, could create materials with self-healing capabilities and stress-stiffening behaviors [108]. These systems would not only maintain structural integrity under dynamic joint loading but also actively promote chondrocyte differentiation through mechanical signaling. Additionally, logic-gate hydrogels integrating multiple stimulus-responsive elements, for example, to pH, ROS, and MMPs, could function as microscopic processors [109]. They would intelligently and sequentially release different therapeutic agents in response to complex pathological signal combinations, thereby achieving precise programmed treatment from anti-inflammatory therapy to pro-regenerative stimulation.

Second, biological control over implanted cell fate remains relatively imprecise. Cells frequently undergo dedifferentiation or terminal differentiation into hypertrophic chondrocytes, leading to calcification of repaired tissue. Future breakthroughs may incorporate synthetic biology principles into tissue engineering. By designing genetic circuits that overexpress anti-hypertrophic factors, such as PTHrP, or engineering synthetic receptors responsive to mechanical stimuli, cells could achieve autonomous decision-making capacity and maintain stable chondrogenic phenotypes. Meanwhile, integrating organoid technology with hydrogels presents an exciting frontier. Pre-forming self-organizing micro-bone or osteochondral organoids in vitro, then implanting them as living building blocks, could simultaneously address multiple challenges, including low cell survival rates, disordered spatial distribution, and poor functional integration.

Finally, we must confront the translational gap directly. Current animal models, predominantly using acute injuries in young specimens, fail to adequately replicate the complex context of human osteoarthritis, characterized by chronic progression, aging, and metabolic abnormalities [110]. Therefore, validating therapeutic efficacy in humanized models or disease-recapitulating models will prove crucial. Moreover, advancing standardized and automated manufacturing, such as GMP-compliant hydrogel production and 3D bioprinting processes, forms the foundation for ensuring treatment reproducibility and safety [111]. This is essential for transitioning these technologies from laboratory prototypes to clinical pharmaceuticals.

In conclusion, this review systematically elaborates on the hierarchical structure and physiological functions of the osteochondral unit, with an in-depth discussion of the pathological vicious cycle formed between cartilage degeneration and abnormal subchondral bone remodeling in osteochondral defects. It further highlights that current pharmacological treatments can only temporarily alleviate symptoms, while surgical interventions often face challenges such as poor tissue compatibility and risks of secondary damage. Against this backdrop, injectable hydrogels and 3D printing technologies demonstrate considerable potential as key tools for precision cartilage repair. Injectable hydrogels exhibit excellent biocompatibility, adapt to irregular defect morphologies after minimally invasive injection, and allow flexible loading of functional components. They enable precise modulation of the injury microenvironment by delivering anti-inflammatory drugs, pro-regenerative factors, stem cells, or therapeutic genes, thereby providing suitable matrix support for cartilage regeneration. Meanwhile, 3D printing technology facilitates the accurate fabrication of biomimetic osteochondral gradient structures through personalized design and multi-material integration. The combination of these two approaches establishes an integrated repair strategy spanning material design, functional loading, and structural biomimicry, opening new avenues for the treatment of complex osteochondral defects.

Declaration of competing interest

The authors declare that they have no known competing financial interests or personal relationships that could have appeared to influence the work reported in this paper.

Acknowledgments

This work was supported by the National Natural Science Foundation of China (82304429), the general funding from China Post-doctoral Funding (2024M751350), the post-doctoral funding from the University of South China (220XQD106), and the Hunan Provincial Natural Science Foundation of China (2023JJ40570).

Author details

Zhi Zheng[1*], Roujun Chen[2] and Yaqin Lin[1]

1 Hunan Province Cooperative Innovation Center for Molecular Target New Drug Study and School of Pharmaceutical Science, Hengyang Medical School, University of South China, Hengyang, China

2 School of Resources, Environment and Safety Engineering, University of South China, Hengyang, China

*Address all correspondence to: zhengzhi@usc.edu.cn

References

[1] Berni M, Marchiori G, Baleani M, Giavaresi G, Lopomo NF. Biomechanics of the human osteochondral Unit: a systematic review. Materials. 2024;**17**:1698

[2] Wang M, Wu Y, Li G, Lin Q, Zhang W, Liu H, Su J. Articular cartilage repair biomaterials: Strategies and applications. Materials Today Bio. 2024;**24**:100948

[3] Hart HF, Crossley KM, Patterson BE, Guermazi A, Birmingham TB, Koskoletos C, Michaud A, De Livera A, Culvenor AG. Adiposity and cartilage lesions following ACL reconstruction. Osteoarthritis and Cartilage. 2024;**32**:931–936

[4] Cross M , Ong KL, Culbreth GT, Steinmetz JD, Cousin E, Lenox H. Global, regional, and national burden of osteoarthritis, 1990-2020 and projections to 2050: A systematic analysis for the Global Burden of Disease Study 2021. The Lancet Rheumatology. 2023;**5**:e508–e522

[5] Honvo G, Leclercq V, Geerinck A, Thomas T, Veronese N, Charles A, Rabenda V, Beaudart C, Cooper C, Reginster J-Y, Bruyère O. Safety of topical non-steroidal anti-inflammatory drugs in osteoarthritis: outcomes of a systematic review and meta-analysis. Drugs & Aging. 2019;**36**:45–64

[6] O'Connor D, Johnston RV, Brignardello-Petersen R, Poolman RW, Cyril S, Vandvik PO, Buchbinder R. Arthroscopic surgery for degenerative knee disease (osteoarthritis including degenerative meniscal tears). The Cochrane Database of Systematic Reviews. 2022;**3**:Cd014328

[7] Thorlund JB, Juhl CB, Roos EM, Lohmander LS. Arthroscopic surgery for degenerative knee: Systematic review and meta-analysis of benefits and harms. BMJ (Clinical Research Ed.). 2015;**350**:h2747

[8] Khan M, Evaniew N, Bedi A, Ayeni OR, Bhandari M. Arthroscopic surgery for degenerative tears of the meniscus: A systematic review and meta-analysis, CMAJ. Canadian Medical Association Journal. 2014; **186**:1057–1064

[9] Sochacki KR, Varshneya K, Calcei JG, Safran MR, Abrams GD, Donahue J, Chu C, Sherman SL. Comparison of autologous chondrocyte implantation and osteochondral allograft transplantation of the knee in a large insurance database: reoperation rate, complications, and cost analysis. Cartilage. 2021;**13**:1187s–1194s

[10] Zheng Z, Sun J, Wang J, He S, Liu Z, Xie J, Yu C-Y, Wei H. Enhancing myocardial infarction treatment through bionic hydrogel-mediated spatial combination therapy via mtDNA-STING crosstalk modulation. Journal of Controlled Release. 2024;**371**:570–587

[11] Zheng Z, Guo Z, Zhong F, Wang B, Liu L, Ma W, Yu C-Y, Wei H. A dual crosslinked hydrogel-mediated integrated peptides and BMSC therapy for myocardial regeneration. Journal of Controlled Release. 2022;**347**: 127–142

[12] Zheng Z, Tan Y, Li Y, Liu Y, Yi G, Yu C-Y, Wei H. Biotherapeutic-loaded injectable hydrogels as a synergistic strategy to support myocardial repair after myocardial infarction. Journal of Controlled Release. 2021;**335**:216–236

[13] Liu Z, Zheng Z, Xie J, Wei H, Yu C-Y. Hydrogel-based cardiac patches for myocardial infarction therapy: Recent advances and challenges. Materials Today Bio. 2024;**29**:101331

[14] Xu B, Ye J, Yuan F-Z, Zhang J-Y, Chen Y-R, Fan B-S, Jiang D, Jiang W-B, Wang X, Yu J-K. Advances of stem cell-laden hydrogels with biomimetic microenvironment for osteochondral repair. Frontiers in Bioengineering and Biotechnology. 2020;**8**:247

[15] Ko C-Y, Ku K-L, Yang S-R, Lin T-Y, Peng S, Peng Y-S, Cheng M-H, Chu I-M. In vitro and in vivo co-culture of chondrocytes and bone marrow stem cells in photocrosslinked PCL-PEG-PCL hydrogels enhances cartilage formation. Journal of Tissue Engineering and Regenerative Medicine. 2016;**10**: E485–e496

[16] Richter DL, Tanksley JA, Miller MD. Osteochondral autograft transplantation: A review of the surgical technique and outcomes. Sports Medicine and Arthroscopy Review. 2016;**24**:74–78

[17] Fang W, Yang M, Wang L, Li W, Liu M, Jin Y, Wang Y, Yang R, Wang Y, Zhang K, Fu Q. Hydrogels for 3D bioprinting in tissue engineering and regenerative medicine: Current progress and challenges. International Journal of Bioprinting. 2023;**9**:759

[18] Zhu W, Mow VC, Koob TJ, Eyre DR. Viscoelastic shear properties of articular cartilage and the effects of glycosidase treatments. Journal of Orthopaedic Research. 1993;**11**: 771–781

[19] Rajagopal K, Ramesh S, Walter NM, Arora A, Katti DS, Madhuri V. In vivo cartilage regeneration in a multi-layered articular cartilage architecture mimicking scaffold. Bone & Joint Research. 2020;**9**:601–612

[20] Zhu M, Zhong W, Cao W, Zhang Q, Wu G. Chondroinductive/chondroconductive peptides and their-functionalized biomaterials for cartilage tissue engineering. Bioactive Materials. 2022;**9**:221–238

[21] Zevenbergen L, Gsell W, Cai L, Chan DD, Famaey N, Vander Sloten J, Himmelreich U, Neu CP, Jonkers I. Cartilage-on-cartilage contact: Effect of compressive loading on tissue deformations and structural integrity of bovine articular cartilage. Osteoarthritis and Cartilage. 2018;**26**:1699–1709

[22] Embree MC, Chen M, Pylawka S, Kong D, Iwaoka GM, Kalajzic I, Yao H, Shi C, Sun D, Sheu T-J, Koslovsky DA, Koch A, Mao JJ. Exploiting endogenous fibrocartilage stem cells to regenerate cartilage and repair joint injury. Nature Communications. 2016;7:13073

[23] Briggs K, Philippon M, Trinidade C, Daniela M. Sport Specific Prevalence of Chondral Injuries in the Hip. British Journal of Sports Medicine. 2017;**51**:302–303

[24] Tong Z, Ma Y, Liang Q, Lei T, Wu H, Zhang X, Chen Y, Pan X, Wang X, Li H, Lin J, Wei W, Teng C. An in situ forming cartilage matrix mimetic hydrogel scavenges ROS and

ameliorates osteoarthritis after superficial cartilage injury. Acta Biomaterialia. 2024;**187**:82–97

[25] Gao L, Cucchiarini M, Madry H. Cyst formation in the subchondral bone following cartilage repair. Clinical and Translational Medicine. 2020;**10**:e248

[26] Reilingh ML, Lambers KTA, Dahmen J, Opdam KTM, Kerkhoffs G. The subchondral bone healing after fixation of an osteochondral talar defect is superior in comparison with microfracture. Knee Surgery, Sports Traumatology, Arthroscopy. 2018;**26**:2177–2182

[27] Reilingh ML, Lambers KTA, Dahmen J, Opdam KTM, Kerkhoffs GMMJ. The subchondral bone healing after fixation of an osteochondral talar defect is superior in comparison with microfracture, knee surgery, sports traumatology. Arthroscopy. 2018;**26**:2177–2182

[28] Hopkins T, Wright KT, Kuiper NJ, Roberts S, Jermin P, Gallacher P, Kuiper JH. An in vitro system to study the effect of subchondral bone health on articular cartilage repair in humans. Cells. 2021;**10**:1903

[29] Donell S. Subchondral bone remodelling in osteoarthritis. EFORT Open Reviews. 2019;**4**:221–229

[30] Grechenig S, Worlicek M, Penzkofer R, Zeman F, Kujat R, Heiss P, Pattappa G, Zellner J, Angele P. Bone block augmentation from the iliac crest for treatment of deep osteochondral defects of the knee resembles biomechanical properties of the subchondral bone, Knee Surgery. Knee Surgery, Sports Traumatology, Arthroscopy. 2019;**27**:2488–2493

[31] Ayala S, Delco ML, Fortier LA, Cohen I, Bonassar LJ. Cartilage articulation exacerbates chondrocyte damage and death after impact injury. Journal of Orthopaedic Research. 2021;**39**:2130–2140

[32] Takahata K, Lin -Y-Y, Osipov B, Arakawa K, Enomoto S, Christiansen BA, Kokubun T. Concurrent joint contact in anterior cruciate ligament injury induces cartilage micro-injury and subchondral bone sclerosis, resulting in knee osteoarthritis. Osteoarthritis and Cartilage. 2025;**S1063-4584**:01058–1

[33] van den Borne MP, Raijmakers NJ, Vanlauwe J, Victor J, de Jong SN, Bellemans J, Saris DB. International cartilage repair society (ICRS) and oswestry macroscopic cartilage evaluation scores validated for use in autologous chondrocyte implantation (ACI) and microfracture. Osteoarthritis and Cartilage. 2007;**15**:1397–1402

[34] Liacini A, Sylvester J, Li WQ, Zafarullah M. Inhibition of interleukin-1-stimulated MAP kinases, activating protein-1 (AP-1) and nuclear factor kappa B (NF-κB) transcription factors down-regulates matrix metalloproteinase gene expression in articular chondrocytes. Matrix Biology. 2002;**21**:251–262

[35] Gómez R, Villalvilla A, Largo R, Gualillo O, Herrero-Beaumont G. TLR4 signalling in osteoarthritis--finding targets for candidate DMOADs. Nature Reviews Rheumatology. 2015;**11**: 159–170

[36] Li M, Nie Y, Zeng Y, Wu Y, Liu Y, Wu L, Xu J, Shen B. Does bisphosphonate increase the sclerosis of tibial subchondral bone in the progression of knee osteoarthritis—a

propensity score matching cohort study based on osteoarthritis initiative. Frontiers in Medicine. 2021;**8**:781219

[37] Huang L, Zhang S, Wu J, Guo B, Gao T, Shah SZA, Huang B, Li Y, Zhu B, Fan J, Wang L, Xiao Y, Liu W, Tian Y, Fang Z, Lv Y, Xie L, Yao S, Ke G, Huang X, Huang Y, Li Y, Jia Y, Li Z, Feng G, Huo Y, Li W, Zhou Q, Hao J, Hu B, Chen H. Immunity-and-matrix-regulatory cells enhance cartilage regeneration for meniscus injuries: A phase I dose-escalation trial. Signal Transduction and Targeted Therapy. 2023;**8**:417

[38] Li M, Yin H, Yan Z, Li H, Wu J, Wang Y, Wei F, Tian G, Ning C, Li H, Gao C, Fu L, Jiang S, Chen M, Sui X, Liu S, Chen Z, Guo Q. The immune microenvironment in cartilage injury and repair. Acta Biomaterialia. 2022;**140**:23–42

[39] Hagen KB, Smedslund G, Moe RH, Grotle M, Kjeken I, Kvien TK. The evidence for non-pharmacological therapy of hand and Hip OA. Nature Reviews Rheumatology. 2009;**5**: 517–519

[40] Bierma-Zeinstra S, van Middelkoop M, Runhaar J, Schiphof D. Nonpharmacological and nonsurgical approaches in OA. Best Practice & Research Clinical Rheumatology. 2020;**34**:101564

[41] Bay-Jensen A-C, Thudium CS, Gualillo O, Mobasheri A. Biochemical marker discovery, testing and evaluation for facilitating OA drug discovery and development. Drug Discovery Today. 2018;**23**:349–358

[42] Orth P, Gao L, Madry H. Microfracture for cartilage repair in the knee: A systematic review of the contemporary literature. Knee Surgery, Sports Traumatology, Arthroscopy. 2020;**28**:670–706

[43] Madry H. Surgical therapy in osteoarthritis. Osteoarthritis and Cartilage. 2022;**30**:1019–1034

[44] Zhou Q, Cai Y, Jiang Y, Lin X. Exosomes in osteoarthritis and cartilage injury: Advanced development and potential therapeutic strategies. International Journal of Biological Sciences. 2020;**16**:1811–1820

[45] Kingsbury SR, Tharmanathan P, Keding A, Watt FE, Scott DL, Roddy E, Birrell F, Arden NK, Bowes M, Arundel C, Watson M, Ronaldson SJ, Hewitt C, Doherty M, Moots RJ, O'Neill TW, Green M, Patel G, Garrood T. Pain Reduction With Oral Methotrexate in Knee Osteoarthritis. Annals of Internal Medicine. 2024;**177**:1145–1156

[46] Simsek I, Swearingen C, Ghandehari H, Kennedy S, Tambiah J, Yazici Y, Skrepnik N. POS0278 A multicenter, observational, extension study evaluating the safety, tolerability, and efficacy of a single lorecivivint injection in knee oa subjects. Annals of the Rheumatic Diseases. 2021;**80**:363–364

[47] Dakin P, DiMartino SJ, Gao H, Maloney J, Kivitz AJ, Schnitzer TJ, Stahl N, Yancopoulos GD, Geba GP. The efficacy, tolerability, and joint safety of fasinumab in osteoarthritis pain: A phase IIb/III double-blind, placebo-controlled, randomized clinical trial. Arthritis & Rheumatology. 2019;**71**:1824–1834

[48] Talouki PY, Tamimi R, Rudi SG. A comprehensive review of curcumin-based scaffolds in cartilage

tissue engineering. Stem Cell Research & Therapy. 2025;**16**:528

[49] He Y, Zisan Z, Lu Z, Zheng L, Zhao J. Bergapten alleviates osteoarthritis by regulating the ANP 32A/ ATM signaling pathway. FEBS Open Bio. 2019;**9**:1144–1152

[50] Tu Y, Yue W, Dai G, Li Y, Liu Y, He C. Anti-resorptive dihydrophenanthrene coelonin alleviates arthritic bone destruction by inhibiting osteoclast formation and synovial inflammation through Src regulation. Phytomedicine. 2025;**146**:157147

[51] Jenei-Lanzl Z, Maurer S, Brenner RE, Zaucke F, Fuchs M, Riegger J. Emerging concepts and challenges in the development of disease-modifying osteoarthritis drugs – A more refined perspective. Archives of Pharmacal Research. 2025;**48**:467–494

[52] Shi Y, Hu X, Cheng J, Zhang X, Zhao F, Shi W, Ren B, Yu H, Yang P, Li Z, Liu Q, Liu Z, Duan X, Fu X, Zhang J, Wang J, Ao Y. A small molecule promotes cartilage extracellular matrix generation and inhibits osteoarthritis development. Nature Communications. 2019;**10**:1914

[53] Geng N, Fan M, Kuang B, Zhang F, Xian M, Deng L, Chen C, Pan Y, Chen J, Feng N, Liang L, Ye Y, Liu K, Li X, Du Y, Guo F. 10-hydroxy-2-decenoic acid prevents osteoarthritis by targeting aspartyl β hydroxylase and inhibiting chondrocyte senescence in male mice preclinically. Nature Communications. 2024;**15**:7712

[54] Jin L, Wang Z, Peng C, He M, Wang F, He H, Liu C. Microneedle assisted melittin-chondroitin sulfate administration for the transdermal therapy of rheumatoid arthritis. Advanced Healthcare Materials. 2025;**14**:2400543

[55] Zheng Z, Yu C, Wei H. Injectable Hydrogels as Three-Dimensional Network Reservoirs for Osteoporosis Treatment. Tissue Engineering. Part B, Reviews. 2021;**27**:430–454

[56] Tan Y, Nie Y, ZhengWen L, Zheng Z. Comparative effectiveness of myocardial patches and intramyocardial injections in treating myocardial infarction with a MitoQ/hydrogel system. Journal of Materials Chemistry B. 2024;**12**:5838–5847

[57] Kureha T, Takahashi K, Kino M, Kida H, Hirayama T. Controlling the mechanical properties of hydrogels via modulating the side-chain length. Soft Matter. 2023;**19**:2878–2882

[58] Zheng Z, Sun J, Wang J, He S, Huang Y, Yang X, Zhao Y, Yu C-Y, Wei H. A negative feedback bionic hydrogel reverses damaged cartilage microenvironment and induces targeted BMSCs differentiation for cartilage regeneration. Chemical Engineering Journal. 2023;**473**:145228

[59] Zheng Z, Lei C, Liu H, Jiang M, Zhou Z, Zhao Y, Yu C-Y, Wei H. A ROS-responsive liposomal composite hydrogel integrating improved mitochondrial function and pro-angiogenesis for efficient treatment of myocardial infarction. Advanced Healthcare Materials. 2022;**11**:2200990

[60] Tan Y, Li J, Nie Y, Zheng Z. Novel approach for cardioprotection: In situ targeting of metformin via conductive hydrogel system. Polymers. 2024;**16**:2226

[61] Gao Y, Guo X, Wang X, Li D, Tang H, Zhao W, Xu T, Wang S-S, Li F,

Zhang H, Feng B, Zhang Y. Engineered mechanical dynamics in biomimetic hydrogel scaffold promotes phenotypic maintenance and cartilage formation of chondrocytes. Composites Part B: Engineering. 2024;**284**:111665

[62] Heng C, Zhou Y, Luo H, Pan H, Cui X, Wei X, Chen L, Xie X. Hydroxyapatite injectable hydrogel with nanozyme activity for improved immunoregulation microenvironment and accelerated osteochondral defects repair via mild photothermal therapy. Biomaterials Advances. 2026;**178**:214462

[63] Yao Z, You X, Wang X, Kang Y, Wang L, Zhang Z. Stem cell-based hydrogel for the repair and regeneration of cartilage. Chinese Chemical Letters. 2025;**36**:110607

[64] Park H, Choi B, Hu J, Lee M. Injectable chitosan hyaluronic acid hydrogels for cartilage tissue engineering. Acta Biomaterialia. 2013;**9**:4779–4786

[65] Zhang T, Huang Y, Gong Y, Shi X, Xiao D, Ren L, Dai X, Zeng Z, Zhao C. A ROS-responsive and scavenging hydrogel for postoperative abdominal adhesion prevention. Acta Biomaterialia. 2024;**184**:98–113

[66] Xiang H, Zhang C, Xiong Y, Wang Y, Pu C, He J, Chen L, Jiang K, Zhao W, Yang H, Wang F, Li Y. MMP13-responsive hydrogel microspheres for osteoarthritis treatment by precise delivery of celecoxib. Materials & Design. 2024;**241**:112966

[67] Deng X, Wu Y, Tang Y, Ge Z, Wang D, Zheng C, Zhao R, Lin W, Wang G. Microenvironment-responsive smart hydrogels with antibacterial activity and immune regulation for accelerating chronic wound healing. Journal of Controlled Release. 2024;**368**:518–532

[68] Xie K, Liu Z, Chen K, Che J, Wang Y. A matrix metalloproteinase-responsive hydrogel with spatially orchestrated delivery of stem cells and raleukin for enhanced bone regeneration. Materials Today Bio. 2025;**35**:102475

[69] Zhu Y, Xiu Z, Jiang X, Zhang H, Li X, Feng Y, Li B, Cai R, Li C, Tao G. Injectable hydrogels with ROS-triggered drug release enable the co-delivery of antibacterial agent and anti-inflammatory nanoparticle for periodontitis treatment. Journal of Nanobiotechnology. 2025;**23**:205

[70] Thomas J, Chopra V, Sharma A, Panwar V, Kaushik S, Rajput S, Mittal M, Guha R, Chattopadhyay N, Ghosh D. An injectable hydrogel having proteogly can-like hierarchical structure supports chondrocytes delivery and chondro genesis. International Journal of Biological Macromolecules. 2021;**190**:474–486

[71] Wei J, Ran P, Li Q, Lu J, Zhao L, Liu Y, Li X. Hierarchically structured injectable hydrogels with loaded cell spheroids for cartilage repairing and osteoarthritis treatment. Chemical Engineering Journal. 2022; **430**:132211

[72] Zha X-J, Ling T-X, Xiao Q, Chen Z-X, Zhang Y, Huang J-G, Zhou Z-K. Integrated 3D printing of topologically hierarchical mechanical hydrogel for accelerating osteochondral regeneration. Bioactive Materials. 2026;**55**:679–692

[73] Guo JL, Kim YS, Koons GL, Lam J, Navara AM, Barrios S, Xie VY,

Watson E, Smith BT, Pearce HA, Orchard EA, van den Beucken JJJP, Jansen JA, Wong ME, Mikos AG. Bilayered, peptide-biofunctionalized hydrogels for in vivo osteochondral tissue repair. Acta Biomaterialia. 2021;**128**:120–129

[74] Chen H, Tao X, Liu J, Kong Q, Wang L, Zhao Y, Jiang C, Song Y, Xu X, Zhu J, Wang Y, Du F, Chen B, Wu J. Inflammatory responsive hydrogels incorporated with bevacizumab-loaded infinite coordination nanopolymers for accelerating cartilage repair. Journal of Colloid and Interface Science. 2025;**700**:138484

[75] Yang R, Yan L, Xu T, Zhang K, Lu X, Xie C, Fu W. Injectable bioadhesive hydrogel as a local nanomedicine depot for targeted regulation of inflammation and ferroptosis in rheumatoid arthritis. Biomaterials. 2024;**311**:122706

[76] Huang K, Zhang Q-Y, Tan J, Nie R, Feng Z-Y, Liu Y, Sheng N, Li H-X, Zhang Y-Q, Shen B, Xie H-Q. Accelerated cartilage regeneration through immunomodulation and enhanced chondrogenesis by an extracellular matrix hydrogel encapsulating Kartogenin. Chemical Engineering Journal. 2024;**497**:154993

[77] Wei X, Liu Y, Liu J, Zhao Z, Cui J, Sang D, Tang Y, Ding Y, Gao T, Xue Z, Lv S, Man Z, Li W, Li N. Injectable Ca2+/Sr2+-crosslinked hydrogel with sustained release of LGK-974 for cartilage repair and osteoarthritis prevention via alleviating inflammatory microenvironment. Materials Today Bio. 2025;**34**:102164

[78] Zhong G, Yao J, Huang X, Luo Y, Wang M, Han J, Chen F, Yu Y. Injectable ECM hydrogel for delivery of BMSCs enabled full-thickness meniscus repair in an orthotopic rat model. Bioactive Materials. 2020;**5**:871–879

[79] Wang B, Wang L, Yuan T, Zhang Y, Yang Q, Ou H, Zhang B, Yang L, Li S. Biomimetic microgrooved methacrylated silk fibroin cartilage scaffold for tracheal injury repair. Advanced Functional Materials. 2025;**39**:e17195

[80] Yang J, Jing X, Wang Z, Liu X, Zhu X, Lei T, Li X, Guo W, Rao H, Chen M, Luan K, Sui X, Wei Y, Liu S, Guo Q. Vitro and in vivo study on an injectable glycol Chitosan/Dibenzaldehyde-terminated polyethylene glycol hydrogel in repairing articular cartilage defects. Frontiers in Bioengineering and Biotechnology. 2021;**9**:607709

[81] Chen J, Zhang Q-Y, Tan J, He T, Qin B-Q, Sheng N, Zhang H, Xie H-Q. Enhanced fibrocartilage regeneration at the tendon-bone interface injury through extracellular matrix hydrogel laden with bFGF-overexpressing human urine-derived stem cells. Chemical Engineering Journal. 2024;**497**:154333

[82] Toh WS, Lee EH, Guo X-M, Chan JKY, Yeow CH, Choo AB, Cao T. Cartilage repair using hyaluronan hydrogel-encapsulated human embryonic stem cell-derived chondrogenic cells. Biomaterials. 2010;**31**:6968–6980

[83] Wei W, Ma Y, Zhang X, Zhou W, Wu H, Zhang J, Lin J, Tang C, Liao Y, Li C, Wang X, Yao X, Koh YW, Huang W, Ouyang H. Biomimetic joint paint for efficient cartilage repair by simultaneously regulating cartilage degeneration and regeneration in pigs. ACS Applied Materials & Interfaces. 2021;**13**:54801–54816

[84] Li W, Shi Z, Jing H, Dou Y, Liu X, Zhang M, Qiu Z, Heger Z, Li N. Streamlined metal-based hydrogel facilitates stem cell differentiation, extracellular matrix homeostasis and cartilage repair in male rats. Nature Communications. 2025;**16**:4344

[85] Zhu J, Yang S, Qi Y, Gong Z, Zhang H, Liang K, Shen P, Huang -Y-Y, Zhang Z, Ye W, Yue L, Fan S, Shen S, Mikos AG, Wang X, Fang X. Stem cell–homing hydrogel-based miR-29b-5p delivery promotes cartilage regeneration by suppressing senescence in an osteoarthritis rat model. Science Advances. 2022;**8**:eabk0011

[86] Long D, Mao G, Li H, Chen H, Huang R, Zhao T, Zhang Z, Zhao X. Injectable nanocomposite hydrogel system for miRNA-based cartilage repair in knee osteoarthritis. Advanced Functional Materials. 2025;**35**:2423227

[87] Zhang Z-J, Hou Y-K, Chen M-W, Yu X-Z, Chen S-Y, Yue Y-R, Guo X-T, Chen J-X, Zhou Q. A pH-responsive metal-organic framework for the co-delivery of HIF-2α siRNA and curcumin for enhanced therapy of osteoarthritis. Journal of Nanobiotechnology. 2023;**21**:18

[88] Ngo TD, Kashani A, Imbalzano G, Nguyen KTQ, Hui D. Additive manufacturing (3D printing): A review of materials, methods, applications and challenges. Composites Part B: Engineering. 2018;**143**:172–196

[89] Jia L, Hua Y, Zeng J, Liu W, Wang D, Zhou G, Liu X, Jiang H. Bioprinting and regeneration of auricular cartilage using a bioactive bioink based on microporous photocrosslinkable acellular cartilage matrix. Bioactive Materials. 2022;**16**:66–81

[90] Fan M, Qiang L, Wang Y, Liu Y, Zhuang H, Guo R, Ben Y, Li Q, Zheng P. 3D bioprinted hydrogel/polymer scaffold with factor delivery and mechanical support for growth plate injury repair. Frontiers in Bioengineering and Biotechnology. 2023;**11**:1210786

[91] Berten-Schunk L, Roger Y, Bunjes H, Hoffmann A. Release of TGF-β3 from surface-modified PCL fiber mats triggers a dose-dependent chondrogenic differentiation of human mesenchymal stromal cells. Pharmaceutics. 2023;**15**:1303

[92] Ashraf S, Park H-K, Park H, Lee S-H. Snapshot of phase transition in thermoresponsive hydrogel PNIPAM: Role in drug delivery and tissue engineering. Macromolecular Research. 2016;**24**:297–304

[93] Yue Y. Designing ultrahigh-water-content, tough, and crack-resistant hydrogels by balancing chemical cross-linking and physical entanglement. ACS Applied Engineering Materials. 2024;**2**:638–648

[94] Baghali M, Ziyadi H, Di Martino A. Engineering smart hydrogels for intelligent drug delivery. Polymer Bulletin. 2025;**82**:2287–2328

[95] Zhang S, Wang H, Liu F, Su Y, Han K, Liu Y, Guan F, Liu H, Ma S. Artificial intelligence-enabled hydrogels: Innovations and applications. Journal of Materials Chemistry B. 2025;**13**:14967–14981

[96] Guo L, Fu Z, Li H, Wei R, Guo J, Wang H, Qi J. Smart hydrogel: A new platform for cancer therapy. Advances in Colloid and Interface Science. 2025;**340**:103470

[97] Xie Z-T, Kang D-H, Matsusaki M. Resolution of 3D bioprinting inside bulk gel and granular gel baths. Soft Matter. 2021;**17**:8769–8785

[98] Zandrini T, Florczak S, Levato R, Ovsianikov A. Breaking the resolution limits of 3D bioprinting: Future opport unities and present challenges. Trends in Biotechnology. 2023;**41**:604–614

[99] Li H, Dai J, Wang Z, Zheng H, Li W, Wang M, Cheng F. Front cover: Digital light processing (DLP)-based (bio)printing strategies for tissue modeling and regeneration. Aggregate. 2023;**4**:e340

[100] Li X, Liu B, Pei B, Chen J, Zhou D, Peng J, Zhang X, Jia W, Xu T. Inkjet bioprinting of biomaterials. Chemical Reviews. 2020;**120**:10793–10833

[101] Wang T, Li Y, Liu J, Fang Y, Guo W, Liu Y, Li X, Li G, Wang X, Zheng Z, Wang X, Kaplan DL. Intraarticularly injectable silk hydrogel microspheres with enhanced mechanical and structural stability to attenuate osteoarthritis. Biomaterials. 2022;**286**:121611

[102] Chen Q, Chen K, Zhang X, Feng H, Feng C, Li X, Zhang D. Recent progress of hydrogel-based biomimetic articular cartilage in mechanical and lubricated enhancement. Friction. 2025;**10**:2223–7690

[103] Li H, Lyu Y, Wang R, Yu H, Lin M, Li Z, Zhong Y, Sheng P, Zhang K, Liao W, Bian L. Supramolecular drug-laden hydrogel based on structural tautomerization enhances drug delivery for rheumatoid arthritis treatment. Bioactive Materials. 2025;**53**:495–506

[104] Cai Y, Fu X, Zhou Y, Lei L, Wang J, Zeng W, Yang Z. A hydrogel system for drug loading toward the synergistic application of reductive/heat-sensitive drugs. Journal of Controlled Release. 2023;**362**:409–424

[105] Lei Y, Wang X, Liao J, Shen J, Li Y, Cai Z, Hu N, Luo X, Cui W, Huang W. Shear-responsive boundary-lubricated hydrogels attenuate osteoarthritis. Bioactive Materials. 2022;**16**:472–484

[106] Lin S-H, Hsu S-H. Smart hydrogels for in situ tissue drug delivery. Journal of Biomedical Science. 2025;**32**:70

[107] Yang M, Liu Y, Duan G, Liang Z, Huang Y, Zhang C, Han X, Ma C, He S, Jiang S. Recent advances of biomass-based smart hydrogel Actuators: A review. Chemical Engineering Journal. 2024;**498**:155157

[108] Wei H, Zhao T, Liu X, Ding Q, Zhang S, Yang J, Wang N, Zhang L, Chai G, Xia G, Li A, Ding C, Liu W. An intelligent controlled release hydrogel with photothermal effect for the treatment of bone defects. Chemical Engineering Journal. 2024;**495**:153597

[109] Culver HR, Clegg JR, Peppas NA. Analyte-responsive hydrogels: intelligent materials for biosensing and drug delivery. Accounts of Chemical Research. 2017;**50**:170–178

[110] Chen C, Wu D, Wang Z, Liu L, He J, Li J, Chu B, Wang S, Yu B, Liu W. Peptide-based hydrogel scaffold facilitates articular cartilage damage repair. ACS Applied Materials & Interfaces. 2024;**16**:11336–11348

[111] Zheng Z, Yu D. Advancement of 3D biofabrication in repairing and regeneration of cartilage defects. Biofabrication. 2025;**17**:2

Chapter 6

Hydrogels and the Brain: Recent Advances in Neuroelectronic Devices

Mahsa Heydari, Elham Farahani, Maryam Shamshiri, Sima Nematipour, Nafiseh Dorosti and Mojtaba Taghipoor

Abstract

Neural signals are transmitted through the brain via electrical impulses and chemical messengers, carrying crucial details about the brain's functional mechanisms. To profoundly understand brain function, neural electrodes are needed to sense these signals. However, traditional rigid electrodes struggle with poor contact on curved surfaces like brain tissue, often causing damage or inflammation during implants. To overcome this challenge, hydrogels, as an effective solution for neuroelectronic applications, are promising due to their biocompatibility in contact with the human body. Hydrogel-based or hydrogel-modified nano/micro-structured electrodes can solve the ineffective contact between soft tissue and rigid electronics, enhancing signal detection and safety. By incorporating conductive materials, such as metal-based nanostructures, conductive polymers, and carbon-based nanomaterials, hydrogel neural electrodes achieve low impedance, high signal fidelity, and long-term stability for neuroelectronic devices. This chapter delves into hydrogel design basics and synthesis routes, from simple mixing to advanced crosslinking. We'll spotlight real-world hydrogel-based neuroelectronic applications in wearable and implantable neural devices, including electrochemical sensors (ECS) for neurotransmitters, microelectrode arrays (MEAs) for electrocorticography (ECoG) and intracortical aims, and EEG electrodes. In addition, we will introduce the future outlook and emerging technologies in the field of neuroelectronics.

Keywords: conductive hydrogels, neuroelectronic devices, neurotransmitters, microelectrode array, EEG electrodes, implantable electrodes, brain-computer interfaces, electrochemical sensors

1. Introduction

Neuroelectronics has stood based on the convergence of neuroscience, biomedical engineering, and materials science, with the aim of creating stable and effective bidirectional interaction between electronic devices and the nervous system. Neuroelectronics enables revolutionary technology in terms of neuromodulatory therapies and brain-machine interfaces (BMIs), with objectives of rehabilitation, modulation, or augmentation of neuronal functions. The inventions such as micro-electrode arrays, electrocorticography, or electroencephalography have led to remarkably detailed information in terms of space and time, helping to understand neuroscience.

Simultaneously, in the area of electrophysiology-based interfaces, the use of electrochemical sensors has enabled the monitoring of neurochemicals, including dopamine, GABA, serotonin, glutamate, or acetylcholine. The above-mentioned sensors are equipped with the capability to monitor, in real-time, the dynamics of the involved neurotransmitters, thereby enabling a clear understanding of the relationship between the two processes – electric events and the release mechanism.

Over the last few decades, neural electrodes have evolved for dual functions of recording and stimulation in neurotherapy, but they are no longer limited to these duties. They are also considered a practical tool for multimodal platforms, enabling simultaneous monitoring of electrophysiological and neurochemical signals. When electrochemical sensors are integrated into EEG or MEA systems, a hybrid platform is formed, capable of correlating changes in electrical signals with corresponding fluctuations in neurotransmitter levels. This integration paves the way into the cognition of neural network behavior and could enable closed-loop neuromodulation systems that respond to both electrical and chemical states of the brain [1, 2].

Michigan- and Utah-type microelectrode arrays, as silicon-based traditional electrodes, have notably advanced electrophysiological studies by facilitating precise, high-resolution recording and stimulation. Despite this, the high rigidity and large stiffness mismatch between hard metals in the GPa range and soft tissues in the kPa range, often leading to immunological reactions, tissue damage, and insufficient signal, posing significant limitations on long-term implantation [2, 3].

To address these challenges, cutting-edge neuroelectronic materials focus on mechanical softness, biocompatibility, and adaptable contact with neural tissues and skin. Hydrogels, as alternative emerging materials, are appealing due to their adjustable mechanical properties and structural correspondence to biological tissues. In conductive hydrogels, the ionic and electrical pathways are merged to enable facile neural signal transmission while preserving tissue flexibility. In contrast to hard interfaces, electronic devices established based on conductive hydrogels limit tissue injuries and chronic inflammations. Moreover, they are capable of supporting consistent long-term electrical recordings and chemical sensing. Their chemical conformability allows for the incorporation of degradable or adhesive properties, which can further improve surgical performance and interfacial durability [4–6].

This chapter is divided into three main sections. First, it introduces the reader to the techniques of hydrogel synthesis, the vital requirements for hydrogels in neuroelectronics, and the fabrication technologies, emphasizing their scalability and compatibility. The second part presents an overview of various non-invasive and invasive neuroelectronic devices, including EEG electrodes, MEAs for ECoG and intracortical aims, along with electrochemical sensors, delving into their operational mechanisms and the issues they face in neural electrical or chemical signal assessment. Finally, the chapter highlights the prospects, as well as limitations and opportunities in the future pathway of neuroelectronics, focusing on long-term applications and multimodal closed-loop systems, which integrate both functions of electrical and chemical detection.

2. Hydrogels

Hydrogels comprise a three-dimensional porous structure, in which hydrophilic polymers are networked together and are capable of absorbing water or biological fluids up to thousands of times their dehydrated weight [6]. The term "network" refers to the presence of crosslinks, which inhibit the water solubility of hydrophilic polymer chains. Upon contact with water, they continue to absorb and swell. The swelling behavior of hydrogels is governed by thermodynamic principles; however, hydrogels remain insoluble in aqueous environments. Variations in the polymer concentration, monomer functionality, structure, and cross-linker used in such gels can modify the hydrogel structure [4].

Self-healing ability, strength, ionic/electronic conductivity, toughness, elasticity, stretchability, adhesiveness, and shape memory are unique physicochemical properties of hydrogels that can be manipulated to meet the requirements of specific applications in diverse fields [7]. Among all remarkable hydrogel characteristics, mechanical compatibility and softness, biocompatibility and prevention of scar formation, ionic conductivity and a suitable chemical environment, as well as the capability for drug delivery to neural tissue, make them appealing in the field of neuroelectronics.

The mechanical properties of hydrogels are influenced by various factors, including the chemical structure and molecular weight of the polymer, polymer crystallization, network topology, type of crosslinking, and crosslink density. Typically, most hydrogels are prepared via random polymerization, which inherently introduces heterogeneity into the crosslinked network, causing mechanical flaws. In addition, the high water content of hydrogels leads to a low polymer chain density per unit volume, which results in weak interactions between adjacent chains and, consequently, poor mechanical properties. The poor mechanical properties of hydrogels severely limit their application as structural materials. For instance, in implantable neural electrodes, the mechanical performance of hydrogels is crucial because they must adjust to match the stiffness of brain tissue (0.1-1kPa) to prevent probe displacement or tissue micromotion-induced damage. The mechanical characteristics of hydrogel structures affect cellular or tissue adhesion properties. In addition, two properties of fatigue resistance and high toughness play a key role in hydrogel-based flexible electronic devices, such as wearable neurotransmitter sensors and flexible neural electrodes. The comfort and adaptability of flexible interfaces with skin tissue are critically important for consistent and reliable monitoring of purposes of interest [8, 9]. In this context, Martinelli et al. fabricated a 3D flower-structured microelectrode array (e-flower) that tackles the challenges of traditional 2D devices in interaction with complex 3D tissues, such as spheroids and organoids. The electrode design was inspired by soft microgrippers, which applied the swelling properties of polyacrylic acid hydrogel grafted onto a polyimide substrate. This e-flower, which has been developed for electrophysiological recording, is capable of achieving curvatures as small as 300 μm within a few minutes [10].

Conductive hydrogels, due to their high water content and insulating polymers, are not inherently suited for neuroelectronics. However, their conductivity can be improved by introducing charge transport pathways through two distinct approaches. The optimal conductivity is typically achieved by integrating both pathways. The two primary pathways for enhancing hydrogel conductivity are ionic and electrical conduction.

In the ionic approach (wet path), charge is carried by the movement of dissolved ions through the hydrogel's water-filled network. The wet path can be created by the incorporation of mobile ions such as Na^+, Cl^-, K^+ into the hydrogel precursor or by applying polyelectrolytes (charged polymers) that comprise fixed ionic groups. For example, a polymer with fixed negative charges (like polystyrene sulfonate (PSS^-) will attract and hold a high concentration of mobile positive ions (cations) from the surrounding solution to maintain charge neutrality. Alginate, chitosan, and hyaluronic acid are all natural polyelectrolytes used to make conductive hydrogels. Conversely, in the electrical approach (wire path), charge is transported via electron flow through a conductive material embedded within the hydrogel. To facilitate electron flow, carbon-based materials (e.g., carbon nanotubes, graphene, carbon nanofibers, etc.), conductive polymers (such as polyaniline (PANI), polypyrrole (PPy), poly(3,4-ethylenedioxythiophene), and poly(ethylene styrene sulfonate) (PEDOT: PSS)), and metal nanoparticles or nanowires (Au, Ag, etc.) are commonly employed. These materials can be added to the hydrogel structure by doping or in-situ polymerization of the hydrogels [11]. Wang et al. introduced a conductive hydrogel-based platform designed for cultivating neuronal cells. The hydrogel matrix is composed of a viscoelastic alginate primary network, which is interconnected by a secondary network formed by the neuronal cell adhesion protein, laminin. Embedded throughout the hydrogel matrix are conductive poly(3,4-ethylenedioxythiophene): polystyrene sulfonate (PEDOT: PSS) particles, serving as an electrically active filler phase. This study, which employed both ionic and electronic pathways to enhance conductivity, demonstrated remarkable stability when used for culturing human-induced pluripotent stem cell-derived cortical neurons over 28 days. The findings revealed that the optical, mechanical, and electrochemical properties of the proposed hydrogel are well-suited for electronic interfaces with brain tissues [12]. **Figure 1** depicts a schematic demonstration of conductive hydrogels' applications in neuroelectronics.

2.1 Synthesis methods

The properties of network configuration, porosity, conductivity, and swelling behavior are influenced by the synthesis technique of the hydrogel. In addition, mechanical strength and biocompatibility are also dependent on the selected synthesis process, making it ideal for a broad spectrum of applications, including drug delivery, tissue engineering, sensors/biosensors, wound dressing, and neural probes [8].

In general, hydrogels can be synthesized in different forms of amorphous (non-crystalline), semicrystalline, and crystalline, based on their supposed applications.

Depending on the technique used in their preparation, hydrogels can be produced in different physical forms and morphologies. Hydrogel particles, 2D layers, and 3D bulk are examples of different forms of synthesized hydrogels [13]. It is noteworthy that both synthetic polymers, such as polyethylene glycol, poly (vinyl alcohol) (PVA), polyacrylamide and natural polymers, including chitosan, alginate, gelatin, and hyaluronic acid, can be used as precursors for hydrogel production. Typically, a necessary component in hydrogel preparation could be a combination of both initiators and monomers. The role of the initiator in the process is very essential since they are responsible for the generation of free radicals in the monomers [14].

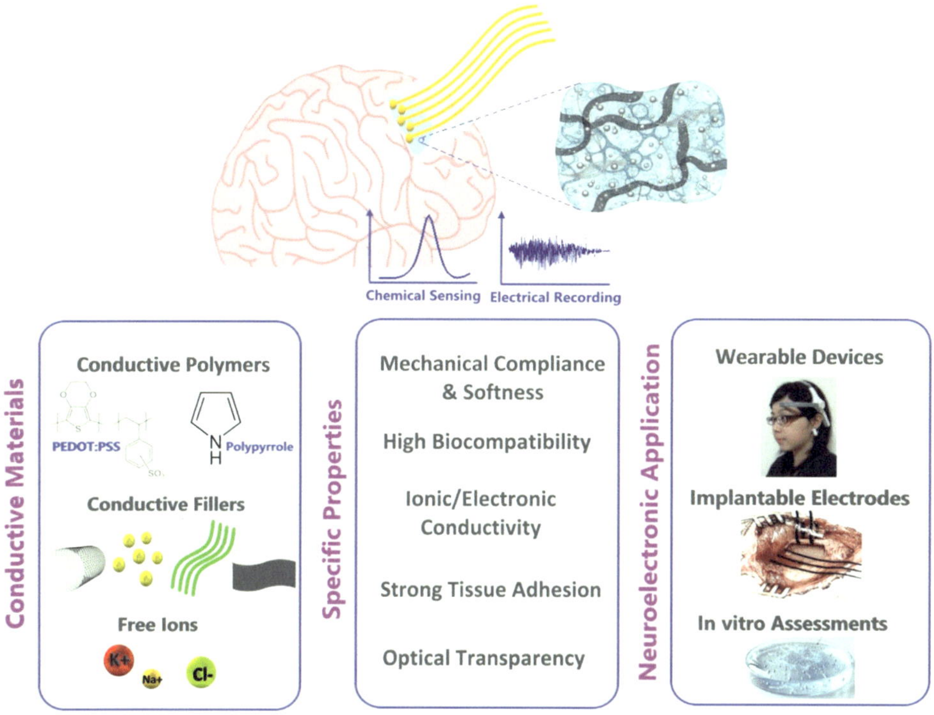

Figure 1.
Summarized characteristics of conductive hydrogels that used in neuroelectronic applications.

Depending on the desired properties, hydrogel formation involves either chemical or physical crosslinking processes that connect polymer chains into a three-dimensional network. In hydrogels, the chemical linkage of polymer chains with an added crosslinking agent affects the physical properties of the polymer depending on the degree of crosslinking and the crystallinity. When the amount of crosslinking agent is insufficient, the physical interaction between polymer bonds breaks easily, leading to the water solubility of the hydrogel. However, excessive use of the crosslinking agent results in a high crosslink density, which reduces the swelling degree of the hydrogel. Crosslinking can make the polymer elastic, decrease viscosity, increase thermal stability, and increase mechanical strength and toughness [15].

Recently, the cutting-edge technologies of the microfluidic method and 3D printing have been integrated into synthesis methods, offering advantages over traditional processes. In the following section, the main different synthesis methods are briefly discussed [16–19].

2.1.1 Physical cross-linking

Physical networks are, in fact, transient cross-links and are created by polymer entanglements due to non-covalent interactions like ionic/electrostatic interaction, hydrogen bonding, polymerized entanglements, hydrophobic/hydrophilic interactions, crystallization/stereo complexation, metal coordination, and π-π

stacking [8, 20, 21]. Unlike conventional chemical cross-linking processes, in this category of hydrogels, the respective interaction is weak and mostly reversible in nature. They are heavily influenced by external factors like pH changes, ionic strength of the solution, solvents, and temperature. This category of hydrogels is stabilized with respect to their structure and simultaneously prevents the complete dissolution of the material. This process of physical cross-linking does not involve harmful initiators and chemicals and hence finds applications in biomedicine and tissue engineering [22, 23].

2.1.2 Chemical cross-linking

Chemical cross-linking refers to the formation of covalent bonds between polymer chains, creating a permanent three-dimensional network. Unlike physical cross-links, these covalent bonds are chemically strong, thermally stable, and irreversible under normal conditions, which is ideal for applications requiring long-term mechanical or chemical stability. The crosslink formation is carried out via thermal, high-energy radiation, crystallite, and enzymatic processes. **Table 1** summarizes the characteristics of the different synthesis methods.

2.1.3 Microfluidic and 3D printing approach

The microfluidic approach allows for precise control of gelation kinetics and morphology by combining polymer precursors and micro/nanostructured agents to create crosslinks in microfluidic chips. Homogeneous microscale hydrogels can be produced with this technique by fine-tuning synthesis parameters like the fraction of chemical ingredients, flow rate, and channel design. It is also possible to create complex and gradient structures using this technology [18].

Hydrogels can also be used as 3D printing ink because of their high hydration level. Hydrogels' adjustable mechanical properties allow for the creation of hydrogels with creative shapes using a variety of 3D printing techniques, including inkjet printing, extrusion, stereolithography, laser-assisted bioprinting, etc. In order to prevent hydrogels from clogging as they exit the printing device's nozzle, the fraction of precursor compounds, viscosity, viscoelasticity, and extrudability of hydrogels must all be optimized. In 3D fabrication, print speed, bed leveling, print orientation, and environmental factors are also crucial. Drug delivery, tissue engineering, and medical devices are just a few of the many uses for hydrogels created using 3D printing [27].

3. Applications of hydrogels in neuroelectronics

Hydrogels can serve as critical interfaces across the diverse spectrum of neuroelectronic technologies, encompassing both non-invasive and invasive methods for neural electrical recording and neurochemical detection.

Neural electrical signals are divided into two categories of field potentials (FP) and action potentials (AP), in which local FPs result from the electrical activities of a group of neurons, and the APs are the consequence of a rapid change in the cell membrane potential. These signals are recorded primarily through three techniques: non-invasive electroencephalography, semi-invasive electrocorticography, and

Method		Mechanism	Advantage	Challenges	Ref
Physical cross-linking	Hydrogen bonding	A type of dipole-dipole attraction when a hydrogen atom bonded to a strongly electronegative atom, exists in the vicinity of another electronegative atom. Hydrogen bonding is essential in the development of polysaccharide-based hydrogels, as the abundant hydrogen donors and acceptors (e.g., OH^-, $COOH^-$, $NH_2{}^-$) present in polysaccharides facilitate the establishment of these interactions.	Safe for clinical purposes Self-healing Injectable properties	Weak interactions Relatively low mechanical strength	[24]
	Ionic interactions	1. Ionic interaction between polyanionic polymers and metal ions. 2. Polyelectroyte complexation of polyanions with polycations.	Safe Self-healing ability	Inherent limitation in mechanical strength	[21]
	Host–Guest interactions	The host is referred to as a molecule that has a large cavity volume, while guest molecules, which have complementary shapes, will interact with the host molecule.	Facile and reversible way	Relatively low mechanical strength	[21]
	Crystallization	Microcrystals can act as join points to form 3D networks. Freeze-thaw technique is a common way to obtain microcrystals.	Simple to operate Safe Enhanced crystallinity and stability	Time-consuming process	[23]

Method		Mechanism	Advantage	Challenges	Ref
Chemical cross-linking	Crosslinking by small molecules	Polymers can be crosslinked using mono/bifunctional small molecules (e.g., formaldehyde, glutaraldehyde, diethyl squarate, etc.) through ionic or covalent interactions.	Forming very strong bonds	Multi-step preparation and purification Cytotoxic for the cells	[25]
	Thermal cross-linking	The heat is applied to make covalent bonds between polymer chains and is applicable to polymer solutions that can tolerate elevated temperatures during processing.	Simple to operate	Extended gelation time. The high temperatures can negatively impact the integrity and viability of biological components.	[8]
	High-energy radiation	The unsaturated reagents can be polymerized through gamma rays or electron beams. The crosslinking degree depends on factors such as polymer content, phase structure, dosage and duration of radiation, as well as the radiation source.	Precise crosslinking Mild aqueous conditions (room temperature, physiological pH) avoid the use of toxic crosslinking agents.	High equipment costs Requires specialized facilities Bioactive substances must be added after irradiation to prevent radical-induced damage.	[25]
	Photopolymerization	Light-sensitive functional groups, such as cyclodextrin, or azobenzene can be a source for hydrogel synthesis.	Speedy preparation Low-cost production	In this method, cell and tissue damages are created due to the increased temperature of light sensitizers and delayed irradiations.	[25]
	Enzymatic cross-linking	Enzyme catalysis modifies the structure and substrates of hydrogels because of the enzymes' chemo-, regio-, and stereo-selectivities.	Most active under mild aqueous reaction conditions. Strongly covalent bonding Rapid gelation (Less than ten minutes)	Substrate specificity The most expensive crosslinkers	[21, 26]

Table 1.
Various methods for synthesizing the hydrogels.

invasive microelectrode arrays. The LFPs are mostly recorded by EEG and ECoG in humans, while APs are usually recorded through animal tests [28]. However, there have also been reports of action potential recordings in humans [29, 30]. Additionally, chemical signals in neural systems have been studied both through in vitro wearable devices and in vivo detections using implanted microelectrodes [31].

As previously discussed, the incorporation of conductive hydrogels in these technologies results in lower impedance, enhanced conductivity, biocompatibility, and superior mechanical properties of the electrodes. In the following sections, a concise overview of each method will be provided, emphasizing the significance of utilizing hydrogels in these domains.

3.1 Electroencephalogram electrodes

EEG is a noninvasive method for recording brain activity by measuring voltage changes on the scalp, forehead, or ear caused by the synchronous activity of neurons. The signal strength obtained from EEG can be considered within a low range (5–300 μV, < 100 Hz). The activities of the brain, which are termed "brain oscillations," have been classified into five categories: alpha, beta, theta, delta, and gamma waves, depending on their frequencies and the activities of the brain at the time of their occurrence. The different frequencies of the electrical activity of the human brain depend on whether the brain is awake, sleeping, or emotionally stimulated [9, 32, 33].

Gamma waves have frequencies ranging from 30–100 Hz and are primarily involved in sensing, remembering, and concentrating or focusing attention on certain tasks. Beta waves have frequencies between 12 and 30 Hz and have been recorded during normal conscious processing of activities and tasks, and are thus attributed to the processes of thinking, concentrated attention, and problem-solving activities of the normal, healthy adult human being. The alpha waves have frequencies ranging from 8–12 Hz and have been recorded while the person is closing his or her eyes, and is relaxed, and at rest, but before the person becomes emotionally disturbed or enters the stages of deep sleep. Theta waves have frequencies of 4–7 Hz and are primarily involved in the process of dozing or transitioning from waking to sleeping, while delta waves have frequencies ranging from 0–4 Hz and are primarily involved in the deep sleep stages of the sleeping pattern of the person [34].

Due to its high temporal resolution, noninvasiveness, and accessibility, EEG allows for a wide range of applications in neurorehabilitation, cognitive neuroscience, and brain-computer interfaces (BCIs).

There are, however, several remaining challenges regarding the attenuation of the signal as it passes through the skull, vulnerability to environmental noise, and conventional electrode limitations; these impose restrictions on achieving consistent and accurate monitoring in ambulatory, wearable, and long-term settings. By ensuring better skin contact and enhanced electrical stability, hydrogel-based electrodes can greatly improve EEG signal quality and stability. They achieve this by reducing skin–electrode contact impedance and minimizing the risk of motion artifacts. Additionally, hydrogels provide a more comfortable experience for long-term use [35].

3.1.1 Classifications of EEG electrodes

One of the main requirements in EEG systems is ensuring the consistency of the electrical link between the subject's scalp area and the EEG electrode. EEG devices can be divided into four main subgroups depending on the method used to make electrical contacts with the scalp, which include wet, dry, semi-dry, and hydrogel-like EEG electrodes, among others, as described briefly below.

3.1.1.1 Wet electrodes

In this type of electrodes, a conductive medium (gel or paste) is inserted between the scalp and the electrodes to create a low-impedance interface between the electrodes and the skin. Two most popular electrodes are the silver/silver chloride (Ag/AgCl) electrodes and the gold cup electrodes, which form the basis of the wet electrodes.

This category possesses a high degree of flexibility and enables high-quality signal recordings. However, the main problem is the application of the gels, which eventually dry out, making long-term data measurement rather problematic. Furthermore, the use of the gels causes skin irritation and sensitivities in the subject [9].

3.1.1.2 Dry electrodes

This class is divided into three subgroups: contact, non-contact, and micro-electromechanical systems (MEMS) electrodes. Unlike wet electrodes, these do not require gel application between surfaces, which reduces allergic reactions and enhances reusability.

In contact-based electrodes, the connection depends on mechanical pressure to establish a lower impedance. The popular comb-structured and claw-shaped designs are used in both commercial products and research. To increase comfort and create softer electrodes, the tips are smoothed by rounding them or incorporating spring-loaded pins.

Non-contact electrodes can record signals without direct physical contact with the skin. In fact, signals are captured through capacitive coupling between the scalp and the electrode. A microstructured fabric, plastic, ceramic, oxide, or organic material is located between the fabricated electrode and the skin surface as a dielectric. Although this method enhances safety and comfort and holds great promise for wearable applications, the low signal-to-noise ratio and high noise necessitate the use of a very high-input-impedance amplifier.

MEMS electrodes are microneedle arrays created through nano/microfabrication processes, including photolithography and micromachining. Due to the penetration of the electrode into the stratum corneum, the outermost layer of the skin, the impedance dramatically reduces; however, the complex and costly fabrication process, along with the discomfort caused by these microneedles, limits their widespread clinical and consumer use [36–38].

3.1.1.3 Semi-dry electrodes

Semi-dry electrodes are designed with a lower need for liquid electrolytes. Furthermore, the benefits of both wet and dry electrodes are combined together in this class. Many attempts have been made to replace aqueous electrolyte-based electrodes with gel-based ones, but good stability has not been reported [9].

3.1.1.4 Hydrogel-like electrodes

Hydrated networks of polymers, such as polyvinyl alcohol (PVA) and polyacrylamide (PAM) with conductive nanomaterial fillers, hold great promise as alternatives to wet Ag/AgCl electrodes, addressing the challenges posed by conventional electrodes [39]. These polymers offer several advantages, including a skin-friendly Young's modulus (1–100 kPa) that enhances user comfort and minimizes motion artifacts [40]. They also provide stable ionic and electronic conductivity, maintaining low impedance (<400 Ω) for extended periods, outperforming wet gels and dry electrodes even during movement or sweating. Furthermore, robust adhesion can be achieved by catechol groups, based on mussel adhesion, as well as nanoclay components, which help in creating a cohesive touch without skin preparation [41]. Many hydrogels also incorporate self-healing polymer chains, preserving performance despite mechanical damage during daily use [42].

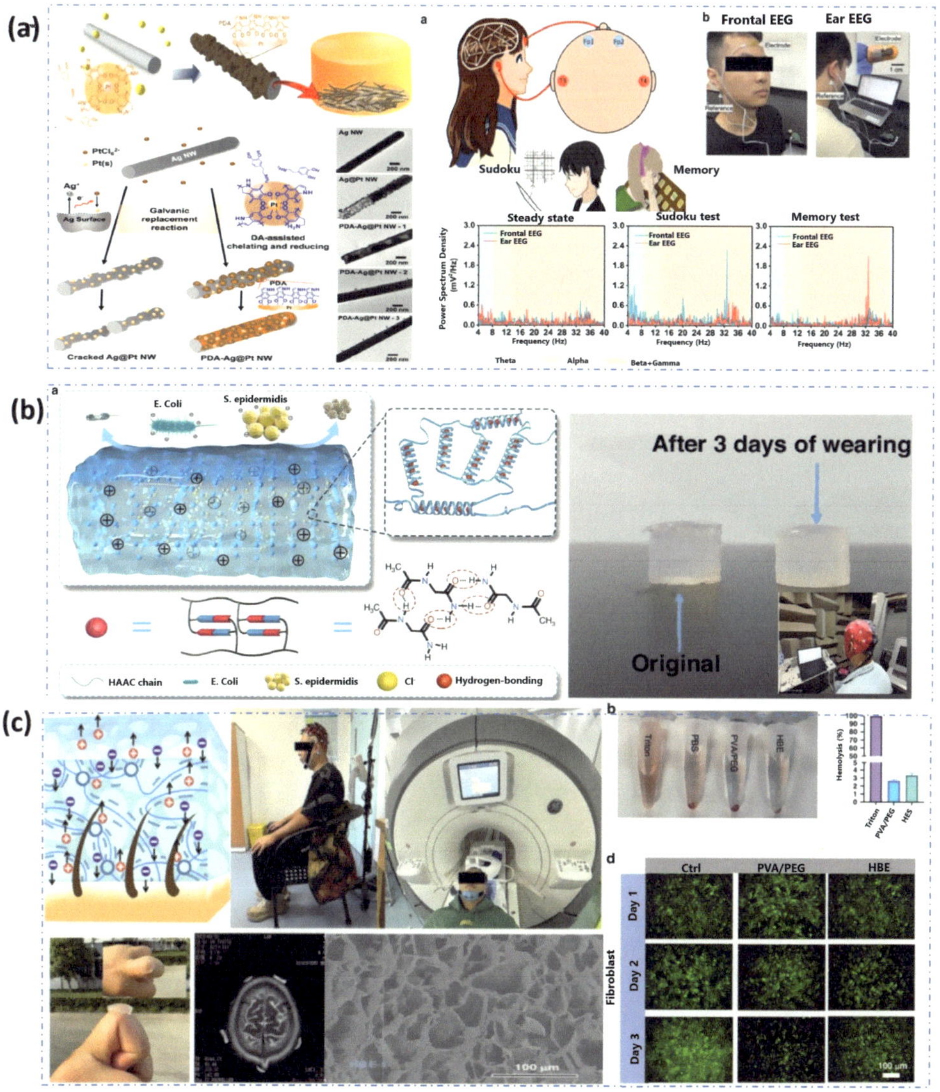

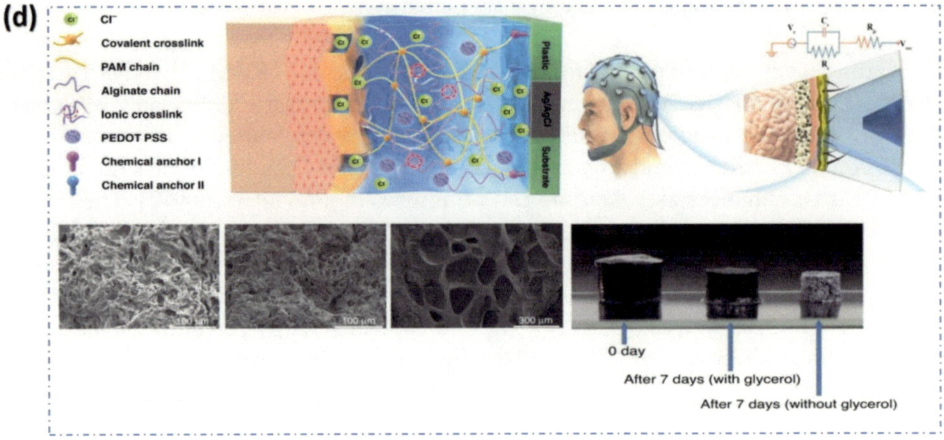

Figure 2.
(a) PDA-PT@ag nanowires formed via pda-assisted galvanic replacement are embedded in a hydrogel to quantitatively analyze high-resolution EEG from the forehead and ear during cognitive tasks, with theta, alpha, gamma and beta waves [40]. (b) The self-repairing hydrogel network, stabilized by hydrogen bonding and ionicinteractions inhibits E. Coli and S. Epidermidis bacteria, maintaining structural integrity during three days use [42]. (c) Hydrogel EEG electrodes record brain activity with two-layer microstructure, maintain stable adhesion under CT/MRI environments. SEM analysis reveals a porous microstructure. Cytocompatibility assessments, including live/dead staining and cell culture imaging, demonstrate minimal cellular toxicity and normal cell morphology after three days, confirming their suitability for long-term bioelectronic applications [43]. (d) Double-layer conductive–adhesive hydrogel captures bioelectric signals through epidermal microchannels, with SEM-confirmed porous structure remaining stable after 7 days in air [35]. (A) (b) (C) Assessment of hydrogel-based electrode in clinical applications. As indicated, the adhesion tests of fabricated hydrogel showed the 90-degree peel adhesion strength [43]. (d) A schematic illustration of fabricated hydrogel-based electrode, and the SEM images of the designed conductive hydrogel [35].

In general, hydrogel electrodes are fabricated as multilayer composites, featuring a conductive layer with materials like graphene, carbon nanotubes, or silver nanowires to enhance electrical performance, alongside an adhesive or bioprotective layer designed for skin attachment and biocompatibility [9]. Recently, Ren et al. reported a hydrogel-based EEG electrode based on PVA and Polyethylene glycol (PEG). By mixing PEG and PVP, this team was able to drastically lower the impedance without using any metal nanoparticles to improve the electrode's conductivity. This allowed electroencephalography to be integrated with MRI and computed tomography techniques (**Figure 2(a)**) [43]. Xue et al. also designed a long-term recording EEG electrode (**Figure 2(b)**), which comprises a double cross-linked network consisting of sodium alginate and polyacrylamide. They used the electrolyte content of water and potassium chloride to enhance the ionic mobility of the hydrogel. The electron conduction within the gel is enhanced by adding the conductive polymers poly (3,4-ethylenedioxythiophene) (PEDOT) and poly(styrene sulfonate) (PSS) to the gel precursor [35]. **Table 2** provides a comparison of the performance across different classes of electrodes.

"Correlation" refers to how similar the signal amplitudes are between two electrode sites, indicating whether two brain regions are active at the same time. In addition, "Coherence" is a more advanced measure that also considers phase, indicating whether two regions are not just active together but are functionally connected and communicating at a specific frequency.

Electrode type	Surface impedance	Signal to noise ratio (dB)	Correlation (%)	Coherence (%)
Wet (Ag/AgCl)	2.8kΩ–130kΩ	–	–	–
Dry (Non-contact)	1MΩ–5MΩ	1.8/3.7	78–92%	60–80%
Dry (Contact)	2.5kΩ–5MΩ	3–24	60–98%	34–90%
Semi-dry	1.5kΩ–61.3kΩ	24.4/24.9	60–98%	80–97%
Hydrogel	0.5kΩ–20kΩ	20–26	70–96%	75–95%
Microneedle	1.181kΩ–2.375MΩ	24/27.1	60–95%	40–86%

Table 2.
A comparison between different electrode types in EEG [44].

3.1.2 Functional merits of hydrogel electrodes

Hydrogel EEG electrodes are well-suited for novel neuromonitoring because of their key advantages, including low impedance, stability, and accurate detection of low-amplitude brain signals, as well as higher signal fidelity during long recording sessions. Their integrated, adhesive, and soft contact surface minimizes motion artifacts and ensures user comfort. These properties ensure that data can be efficiently acquired while performing routine activities and during sleep recording. Additionally, the breathability of the biocompatible polymer network eliminates the risk of irritating the skin and allows for long-term use of the EEG sensor, which is ideal in pediatric and elderly patients and in designing wearable EEG sensors. By combining comfort, stability, and high signal quality, hydrogel electrodes offer clinical-grade accuracy for continuous and mobile brain monitoring [35, 44].

3.1.3 Wireless and wearable hydrogel EEG electrodes

The integration of hydrogel electrodes with wireless wearable electronics is revolutionizing the diagnosis and monitoring of neurological and behavioral disorders. Wireless hydrogel EEG patches maintain reliable signal quality during both activity and sleep. These wearable electrodes exhibit excellent stretchability, fatigue resistance, and operational flexibility across a wide temperature range. Integration with Bluetooth or Near-Field Communication (NFC) chipsets enables real-time data transmission to cloud-based or personal devices, supporting continuous health monitoring. Hydrogel arrays are now being applied in high-density scalp and ear EEG systems across medical, consumer, and research settings [45, 46]. For example, Han and colleagues fabricated a flexible, self-adhesive hydrogel-based EEG electrode that exhibits low skin–electrode impedance, high signal stability, and strong resistance to motion and sweat, enabling reliable wireless EEG recording over extended periods while maintaining user comfort [47].

3.1.4 Hybrid functionality

Next-generation EEG electrodes use microfluidic technologies for hydration management, long-term electrode wear, and simultaneous chemical sampling (e.g., sweat analysis or drug delivery). Microchannels distribute fluids or functional agents throughout the patch, enabling water and ion replenishment to maintain optimal impedance levels.

Recently, 3D printing has facilitated the rapid prototyping of hydrogel electrodes with customized geometries, leading to fatigue-resistant and personalized neural interfaces. Gencai et al. developed a fully flexible hydrogel electrode for daily EEG monitoring, composed of materials that are flexible and conform to the curvature of the skin [35, 48]. The incorporation of glycerol into the N-acryloyl glycinamide (NAGA) hydrogel enhanced its moisturizing properties, allowing it to retain its elasticity even after 90 days of exposure to air [49].

3.1.5 Future directions in EEG: Multimodal smart biosensors and BCI applications

Hydrogel-based patches enable the integration of electrical EEG electrodes with chemical sensors, drug-delivery, and other modalities. These patches are used for continuous cognitive monitoring and biobehavioral analysis in settings like workplace neuroergonomics and education, and they also support brain–computer interfaces for rehabilitation, prosthetics, and neurofeedback [50, 51].

In recent years, several studies have been conducted to find a meaningful relationship between EEG signals and biochemical data, including sweat glucose, lactate, and pH, for personalized medical diagnostics. For example, Chen et al. reported the development of a 200 μm microelectrode for the simultaneous detection of EEG signals and dopamine, one of the most crucial neurotransmitters. The optimized electrode was fabricated based on a combination of boron-nitrogen co-doped graphene–diamond microelectrode. The research team demonstrated that the brain signals alpha, beta, and theta, each of which has varying strengths in various states of brain function, are correlated with dopamine. They discovered that the intensity of beta waves, which are indicative of brain activity, increased in the rats' brains following dopamine injections. They demonstrated the applicability of the developed electrode in remote online diagnosis and treatment of neurological disorders [31].

Hydrogel-based EEG electrodes have the potential to advance noninvasive neuroelectronics and solve many challenges, such as signal quality, user comfort, and long-term monitoring, due to their good skin adaptability and self-healing capabilities. These features make them suitable not only for clinical applications but also for real-world scenarios that require continuous monitoring of physiological, cognitive, and behavioral activities. By bridging the gap between ionic and electronic conductivity and adapting to dynamic physiological and chemical states, hydrogel-based EEG interfaces serve as catalysts for the development of multifunctional biosensors and brain–machine interaction.

3.2 MEAs for electrocorticography and intracortical aims

In addition to EEG, magnetoencephalography (MEG) and functional near-infrared spectroscopy (fNIRS) have also been applied for evolution of non-invasive neuroelectronic platforms. However, non-invasive neural devices face several drawbacks in numerous applications and are not sufficiently suitable for complex tasks, such as the control of robots.

For recording electrical and chemical neural signals, implantable electrodes are of great interest due to their ability to establish a two-way connection (recording and stimulation) with the brain, as well as their high potential for both diagnosis and treatment in various sectors of neuroscience and biomedicine.

Over the past few decades, microelectrode arrays have become integral to the field of neuroscience. Depending on their application, they are typically implanted in various locations, such as the peripheral nerve, cerebral cortex, or spinal cord. Within the central nervous system, they serve multiple clinical purposes, including the treatment of epileptic seizures and the alleviation of Parkinson's disease. Moreover, in the peripheral nervous system, they can function as controllers for neural prostheses, aiding in the restoration of hearing or vision disorders. Moreover, MEAs are commonly used as cell culture mediums for in vitro studies, such as drug screening and assessment of neural network dynamics [1, 8, 51].

3.2.1 Working mechanism

Electrocorticographic recordings are taken from the superficial cortex of the brain. Therefore, due to the omission of the skull bone effect, ECoG signals are recorded with weaker noise and higher bandwidth compared to EEG signals.

The ECoG recording can be performed through two approaches: epidural and subdural. The latter is more intrusive than the epidural approach due to the different surgical methods. While during the epidural method no tissue scarring occurs, the dura mater is damaged in the subdural method. Still, the damage is strongly minimal compared to invasive intracortical MEAs. The ECoG signal strength, recorded from the brain's epicortical area, is in the medium range (0.01–5 mV, < 200 Hz).

In contrast, intracortical electrodes can be directly implanted into the cerebral cortex to capture neural signals produced by neurons, allowing for the examination of neural activity or interaction with external systems. These microelectrodes are inserted into the brain using needle-shaped electrodes to record the APs (about 500 μV, 0.1–7 kHz) and LFPs (<1 mV, < 200 Hz) [52].

3.2.2 Fabrication methods

Upon implantation of electrodes, the neural electrode is considered a foreign object by the body, which causes the stimulation of the immune system and, subsequently, the inflammatory response of the brain to counteract the foreign body's presence. To minimize damage and maximize interaction between the electrodes and the tissue, it is vital to utilize structures representing values of Young's modulus of the brain. Consequently, the development of neural electrodes is shifting from the utilization of stiff materials like platinum, iridium oxide, titanium nitride, silicon, and tungsten to soft substrates, including polyimide, parylene, PDMS, cellulose nanocomposites, biodegradable materials, and shape memory materials [28].

A microelectrode, as a neural interface, is fabricated using a rigid or soft substrate, as previously described. In addition to choosing the appropriate material that is highly conductive and compatible with biological systems, the shape and dimensions of the microelectrode matter significantly. Although the tip of the sensor electrode may be either square or circular, the circular surface is preferred due to the stronger electric field. Usually, there are three basic components found in the structure of a neural interface electrode. These are the substrate, the conductive material, and the insulating layer. This coating ensures that only selected areas of the electrode make contact with the tissue, while the rest remains isolated.

Microfabrication techniques are pivotal in developing high-resolution neural interfaces that can interact with neural tissues at the microscale. By employing methods such as photolithography, laser patterning, screen printing, electrical discharge machining, stereolithography, complementary metal-oxide-semiconductor techniques, and inkjet printing, researchers have fabricated a wide spectrum of 2D and 3D microelectrode arrays that offer precise spatial resolution and adaptability to neural structures [53]. **Figure 3** presents rigid and soft MEAs fabricated through different techniques.

An outstanding report is the study by Shen et al., who developed microfabricated intracortical extracellular matrix-microelectrodes. These electrodes, fabricated on natural-substrate-based organic materials, aimed to minimize the introduction of non-natural products into the brain, enhancing biocompatibility and reducing immune response [54].

3.2.3 Hydrogel-based MEAs

In light of the challenges posed by rigid electrodes, different approaches have been taken to address this problem in the field of microelectrode array fabrication. One of these approaches is the application of low modulus plastic or elastomeric materials. However, these materials also cannot mimic the softness of brain tissue.

Fabrication of MEAs, combined or modified with hydrogels, is another promising approach that promises to provide greater biocompatibility, less invasive behavior, and greater similarity to brain tissue [60–62].

Among all reported works, PEDOT:PSS-based conductive hydrogels are highly promising because of their high biocompatibility, mechanical softness, and electrical conductivity, which are crucial for efficient neural signal transmission. Recent technological breakthroughs have demonstrated the successful interfacing of PEDOT: PSS hydrogels in 3D-printing applications. For example, the printing of PEDOT: PSS was highlighted in a study by Xie et al., where hydrogels were fabricated using a liquid-in-liquid printing method, achieving structures with adjustable mechanical properties for implantation purposes. Similarly, Wang et al. designed 3D-printed PEDOT: PSS neural electrodes capable of maintaining stable electrophysiological recordings in vivo, underscoring the potential of these materials for chronic neural interfacing [63]. These developments emphasize the possibility of utilizing 3D-printed conductive hydrogels as the foundation for creating neural interfaces that not only mimic the mechanical properties of neural tissues but also support and facilitate electrical communication [64]. Moreover, Athukorala et al. have investigated state-of-the-art 3D-printed electroconductive hydrogels, consisting of conductive polymers such as polyaniline and polypyrrole, focusing on their applications in neural tissues. The combination of microfabrication technology with hydrogel materials has enabled the development of neural interfaces that are not only biocompatible with neuronal tissue but also capable of providing high-quality electrical recordings and stimulation. These interfaces are essential for advancing our understanding of neural functions and for developing therapeutic strategies for neurological disorders [65].

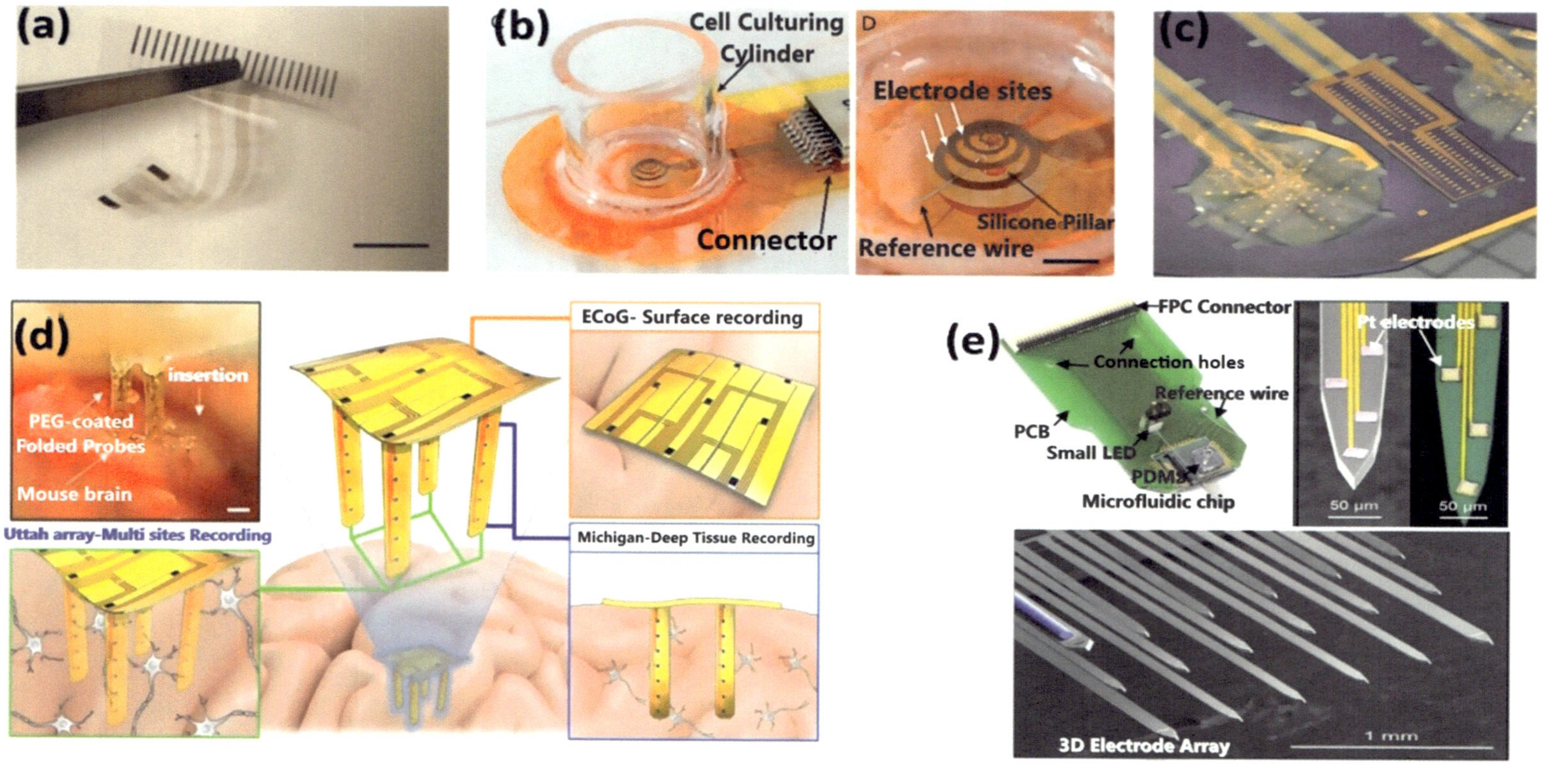

Figure 3.
(A) A 22-channel ECoG MEA is inkjet-printed on a thin flexible parylene film. The gold sensing areas are arranged in a 4 × 5 matrix, each measuring 100 μm by 100 μm [55]. (B) A novel 3D spiral microelectrode array with 50 μm diameter recording sites for in vitro extracellular action potential recording [56]. (C) A designed 4" wafer for visual cortical interfacing, which includes three flexible-rigid ECoG devices, standing free. The polyimide was fully processed on the wafer surface by a combination of dry etching, metallization and Deep Reactive Ion Etching (DRIE) methods. The areas designated for signal acquisition are extremely flexible, while the islands for connector assembly remain rigid [57]. (D) A foldable 3D neural electrode array for simultaneous signal recording of the cortical surface and intracortical area. This device was fabricated based on the integration of ECoG, Utah array, and Michigan probe designs. Each electrode has a recording area of 20 × 20 μm^2 with 500 μm intervals for the ECoG electrodes and 200 μm intervals for the probe electrodes [58]. (E) A packaged 3D MEA with a PDMS fluidic interface. This device is integrated with a light-emitting diode (LED) and a flexible printed circuit connector on a printed circuit board (PCB) for multifunctional applications, including optical stimulation and drug delivery for assessment of neural circuit mechanism [59].

As mentioned above, microelectrode arrays enable two-way communication with the brain through recording and stimulation. Brain stimulation can be performed electrically, optically, or chemically and has the potential to control or treat many diseases related to neurological disorders such as Parkinson's or epilepsy. Recently, Jiang et al. designed and introduced a multifunctional hydrogel-based fiber for neural sensing and deep brain manipulation based on photodynamic therapy. This electrode effectively mediated optogenetic brain stimulation and was able to achieve a signal-to-noise ratio of 10. The fabricated electrode, which took advantage of the ionic chelation/dichelite effects, also achieved a low bending stiffness of 0.3 N/m and a high conductivity of about 97 S/m at 1 kHz. This study, which successfully investigated the behavior of hippocampal neurons in rats in order to control epilepsy, provided an innovative approach and interesting insights into brain circuits and their connection to external electrical systems [69]. **Figure 4** shows some examples of hydrogel-based MEAs, developed for electrophysiological assessments.

3.3 Electrochemical sensors for chemical language of the brain

Neurotransmitters, the brain's chemical signals, are endogenous chemicals crucial for numerous brain functions, including thinking, decision-making, breathing, heart rate, sleeping, speaking, and even emotions and perception. Alterations in the concentration of these chemical messengers in the brain and body can lead to mental disorders such as Alzheimer's, Parkinson's, epilepsy, cardiac arrhythmia, dementia, and addiction. According to the World Health Organization, one-third of the global community suffers from neurological disorders. Consequently, there is an increasing demand for a tool capable of accurately measuring changes in neurotransmitter concentrations to predict or manage these diseases [43].

To date, various methods have been used to detect neurotransmitters. These involve analytical methods such as HPLC, optical methods (e.g., Raman spectroscopy, fluorescence, calorimetry, etc.), as well as microdialysis. However, these methods are performed by specialized personnel and take a considerable amount of time. Thus, these methods cannot be ideally utilized for point-of-care evaluation. Furthermore, owing to their usage of complicated equipment, these methods can be deemed expensive. Meanwhile, the ECSs have received considerable attention owing to the wide advantages these systems hold. These include being inexpensive, ideal systems for point-of-care evaluation tests, and providing fast results. Furthermore, the miniaturization and inclusion capability into the nerve electrical signal electrodes make them a considerable asset in the field of neuroscience [70, 71].

3.3.1 Working mechanism of ECSs

The applications for these sensors lie in areas including the control of food quality, environmental analysis, safety in industries, and medical diagnosis. The remarkable properties, including high sensitivity, fast response time, high selectivity, ultralow limit of detection, cost-effectiveness, high signal-to-noise ratio, and miniaturization, have made them valuable and efficient for the development of advanced technologies such as neuroelectronics and BCI [31, 70].

DOI: http://dx.doi.org/10.5772/intechopen.1014533

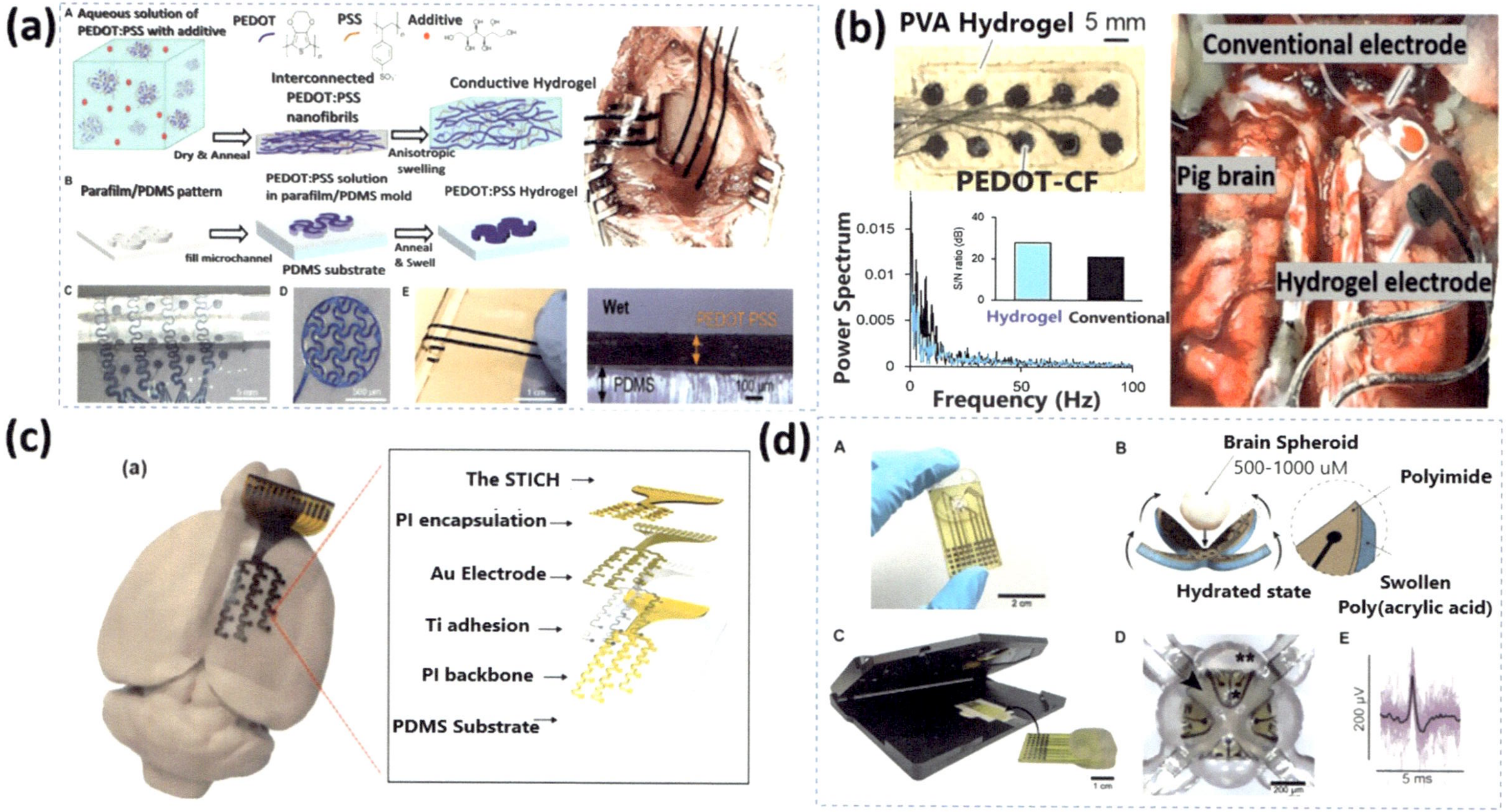

Figure 4.
(a) Schematic illustration of synthesizing the PEDOT:PSS conductive hydrogel and patterning it on a micromolding-fabricated PDMS thin film. It shows the mesh-structured electrode array, which covers a glass rod and an individual electrode [66]. (b) Optical image of a PVA hydrogel-based electrode, which applied as ECoG monitoring on the porcine brain. The recorded brain waves can be seen [67]. (c) A stretchable ECoG electrode array, which uses the soft tough ionic conductive hydrogel (STICH), deposited on the metal electrode. The STICH is comprised of alginate, carboxymethyl cellulose, PEDOT:PSS and polyacrylamide [68]. (d) Optical image of the e-flower system, which consists of a flower-shaped polyimide substrate, each hosting eight platinum MEAs. These Pt MEAs are covered with a layer of polyacrylic acid (PAA) hydrogel [10].

An electrochemical sensor is a device that identifies chemical compounds by their electrochemical characteristics, such as electron transfer, ion exchange, or oxidation-reduction reactions. The chemical data are transformed into an electrical signal, which is then processed and analyzed to ascertain the concentration of the target materials [72].

Electrochemical sensors with two or three-electrode configurations are powerful tools for detecting and measuring a wide range of chemical compounds. An electrochemical sensor with a three-electrode structure consists of a reference electrode that provides a constant potential for comparison, a counter electrode that acts as a current balancer in the system, and a working electrode where chemical reactions take place. The working electrode is a conductive material that is usually surface-modified with conductive nanomaterials such as carbon-based materials (e.g., graphene, carbon nanotubes, carbon quantum dots, etc.) and non-carbon materials (e.g., metal nanoparticles/nanowires, MXenes, metal oxides, etc.), as well as conductive polymers with various dimensions and geometries, in order to increase conductivity and create a larger effective specific surface area.

To fabricate a biosensor, biorecognition components such as aptamers, nucleic acids, proteins, cells, or enzymes are needed to be immobilized on the electrode surface, significantly boosting the electrode's selectivity. Electrochemical biosensors are more selective than ECSs because biorecognition elements have a greater affinity for interacting with analytes of interest, such as ions (e.g., Na^+, Cl^-), small molecules (e.g., dopamine, glutamate, and norepinephrine), microbes, and other species [73, 74].

An electrical signal is produced when biorecognition elements or nanomaterials placed on the electrode surface are correctly attached to the target material (analyte). To translate chemical interactions into electrical signals, electrochemical transducers use a variety of techniques, including conductometry, ampacity, amperometry, voltammetry, potentiometry, and cutting-edge technologies like photoelectrochemistry (PEC), electrochemiluminescence (ECL), and organic electrochemical transistors (OECT). The generated electrical signal is then transformed by the signal processing system into comprehensible data and a visual display for the user. This enables sensors and biosensors to function well in a variety of analytical applications [75].

3.3.2 Electrochemical measurement techniques

As previously stated, the basis of electrochemical techniques is the measurement of variations in conductivity, impedance, potential, and current between two electrodes. These methods fall into two groups: dynamic (with current flow) and static (with zero current). Static techniques, which include direct potentiometry and potentiometric titration, are performed without applying an external current. In contrast, dynamic techniques actively control the electrical environment and are categorized as either potentiostatic (with potential control) or galvanostatic (with controlled current).

The subgroups of potential control methods include voltammetry, amperometry, and polarography.

Among all measurement techniques, potentiometry, amperometry, voltammetry, and impedimetry are the most commonly used methods, which are briefly discussed below.

3.3.2.1 Potentiometric sensors

This class of sensors calculate the changes in electrical potential caused by the activity of ions. This is used in pH sensors and ion-selective electrodes (ISE) to detect specific types of ions, such as sodium, potassium, and chloride.

3.3.2.2 Amperometry sensors

These sensors determine the current produced by an oxidation or reduction process at a steady potential. The basis of amperometry lies in the relationship between the diffusion current intensity and the concentration of the electroactive compound used in biomarker sensors and gas detection.

3.3.2.3 Voltametric sensors

This group uses a variable voltage and measures the current that follows to obtain electrochemical data. The most important voltammetry methods include cyclic voltammetry (CV), linear sweep voltammetry (LSV), normal pulse voltammetry (NPV), differential pulse voltammetry (DPV), square wave voltammetry (SWV), anodic stripping voltammetry (ASV), cathodic stripping voltammetry (CSV), and chronoamperometry (CHA). This method is based on polarization and is applied to drug testing, biochemical research, and heavy metal identification.

3.3.2.4 Impedimetric sensors

This method is an effective technique for studying the rate of electron transfer and diffusion in electrochemical reactions. The spectrum is then plotted in terms of imaginary resistance versus real resistance, which is known as a Nyquist plot. The semicircle corresponding to high frequencies relates to the electrochemical process of electron transfer, while the linear portion at low frequencies relates to the diffusion process. The conductivity technique is a subset of the impedance spectroscopy technique [76].

3.3.3 Hydrogel-based electrochemical sensors for neural chemical signals

The use of hydrogels in electrochemical sensors for the sensing of neurotransmitters is an important and fundamental improvement over conventional rigid electrodes. These advantages directly improve the sensor functionality and the signals quality that can be recorded. Hydrogels serve as a soft cushion, which is capable of evenly distributing pressure, making them particularly suitable for wearable and implantable sensing applications.

For wearable applications, these sensors are integrated into flexible patches or tattoos to non-invasively monitor neurotransmitters or their metabolites in biofluids like sweat or tears. This enables continuous, real-time tracking of mental states, stress levels, or neurological health markers outside a clinical setting. Recently, Liu et al. reported a flexible hydrogel-based sensor for the detection of dopamine. The soft electrode was designed based on the integration of acrylamide (AAM), carbon nanotubes (CNTs), and MoS_2. The synthesized hydrogel showed an excellent limit of detection (LOD) of 6.1 nM in the presence of high concentrations of epinephrine.

Moreover, the mechanical durability of the hydrogel was maintained even after 15 repeated stretch–release cycles (0–50%) [77].

For implantable applications, the hydrogel's softness and biocompatibility are paramount, as they minimize the chronic foreign body response that typically degrades the performance of rigid probes. Implanted hydrogel sensors can provide high-fidelity, real-time data from deep brain structures, allowing for direct investigation of neurochemical dynamics during behavior or in response to treatment. The hydrogel's high ionic conductivity ensures a low-impedance interface, which is crucial for obtaining a clean, high-signal-to-noise ratio, making them vastly superior to traditional electrodes. Furthermore, the hydrogel can be engineered to be self-healing or to deliver therapeutic drugs, creating a smart interface that can both sense and treat neurological conditions, paving the way for personalized neuro-medicine and advanced brain-computer interfaces. Gu et al. designed a biocompatible conductive hydrogel for in vivo dopamine monitoring. The conductivity of the hydrogel was enhanced by polydopamine, $AgInS_2$ and PEDOT: PSS filler, and synthesized through copolymerization of acrylamide and PR-PEGMA crosslinker. The hydrogel-based microelectrode exhibits tissue-matching elasticity (Young's modulus ≈ 118 kPa) and high conductivity (177 mS m^{-1}). This work is highly promising for the development of minimally invasive, high-sensitivity tools for brain research [78]. In other work, Wu et al. introduced aptamer-graphene microtransistors (AGMs) as a breakthrough platform for long-term, real-time dopamine monitoring in vivo, addressing key limitations of prior aptamer-based sensors that degrade within hours due to biofouling and nuclease activity. AGMs feature graphene microtransistors functionalized with aptamers, coated with pyrene-PEG5-alcohol and DNase inhibitor-doped polyacrylamide hydrogel to enhance stability [79].

4. Summary and outlook

In the flourishing field of neuroelectronic devices, hydrogels have emerged as a compelling bridge between rigid electrodes and soft biological tissue. Their unique combination of tissue-mimicking mechanical compliance, ionic-electronic conductivity, and capacity for functionalization opens new routes for long-term neural interfacing.

Compared to traditional rigid silicon or metal microelectrodes, which often suffer from chronic immune responses, micromotion damage, and signal degradation due to mechanical mismatch, hydrogel-based chemical and electrical neural devices promise reduced foreign-body reactions and improved stability. In addition, wearable EEG and electrochemical sensors provide comfort for long-term brain monitoring.

The EEG electrodes are practical for high-quality recording of brain waves and can be used for prolonged monitoring of neurological disorders such as epilepsy. Although EEG electrodes are capable of monitoring psychophysiological conditions and mental functions in patients, more accurate monitoring of brain activity requires the implantation of microelectrodes within the brain. The microelectrodes that are capable of monitoring and stimulating the brain are highly promising and can be used to treat patients with Parkinson's disease, epilepsy, and motor disorders.

In addition, electrochemical sensors enable either invasive or non-invasive monitoring of the most important neurotransmitters and their metabolic products in biofluids such as sweat or blood. This paves the way for the continuous monitoring of biomarkers for psychological conditions, including dopamine, serotonin, glutamate, etc. It appears that, through their combination with wearable EEG electrodes or implantable microelectrodes as multi-modal systems, electrochemical sensors can be a most beneficial element in their own right in enhancing our understanding of brain functions in general. This would give a truly holistic perspective regarding activities in the brain, in combination with its electrical symphony and chemical messages, in a way that has not been seen or fully comprehended before.

Future advances in neuroelectronics are expected to lead to the development of ultra-flexible electrodes with very small dimensions and high densities for use in confined spaces, such as the brain. This approach reduces tissue damage and improves spatial resolution.

Another effective approach in this area is the use of wireless connectivity for signal recording and measurement, in which electronic components such as multiplexers, amplifiers, and signal processors are integrated on a chip, reducing the complexity of external recording systems and enabling real-time processing for wearable or implantable devices. These systems also support real-time and long-term data recording during the natural movement of the subject (human or animal) and reduce energy consumption due to the use of internal rechargeable batteries.

It appears that in multi-modal systems, there is a need for microfluidic channels embedded in microelectrode arrays. In the case of systems that will simultaneously measure reliable electrical and chemical signals, it will be possible to inject drugs to modulate neural activity through these channels.

Given the rapid growth of artificial intelligence technology, another approach will be the use of algorithms to analyze data and decode electrical and chemical signals. Also, the application of closed-loop neuroelectronic systems makes neurophysiological monitoring and intervention possible in real time. By using a responsive neural stimulation in these systems, epileptic activity will be detected, and targeted stimulation will be provided to prevent seizures. This approach is also very promising for the treatment and management of Parkinson's disease through motor neuroprosthetics.

Although the challenges in translating this technology into clinical applications have not been completely resolved, given the significant advances in this field, it seems that personalized treatment in the field of neurological diseases will be achieved in the future.

Conflict of Interest

The authors declare no conflict of interest.

Author details

Mahsa Heydari[1*], Elham Farahani[1], Maryam Shamshiri[2], Sima Nematipour[1], Nafiseh Dorosti[2] and Mojtaba Taghipoor[1,3*]

1 Institute for Convergence Science and Technology, Sharif University of Technology, Tehran, Iran

2 Department of Medical Physics and Biomedical Engineering, School of Medicine, Shahid Beheshti University of Medical Sciences, Tehran, Iran

3 Department of Mechanical Engineering, Sharif University of Technology, Tehran, Iran

*Address all correspondence to: mtaghipoor@sharif.edu; mah.heydari@sharif.edu

References

[1] Duan W, Robles UA, Poole-Warren L, Esrafilzadeh D. Bioelectronic neural interfaces: Improving neuromodulation through organic conductive coatings. Advanced Science. 2024;**11**(2306275). DOI: 10.1002/advs.202306275

[2] Yang D, Tian G, Chen J, Liu Y, Fatima E, Qiu J, et al. Neural Electrodes for Brain-computer Interface System: From Rigid to Soft. Vol. **3**; John Wiley and Sons Inc: BMEM; 2025

[3] Boufidis D, Garg R, Angelopoulos E, Cullen DK, Vitale F. Bio-inspired electronics: Soft, biohybrid, and "living" neural interfaces. Nature Communications. Nature Research. 2025;**16**:1861. DOI: 10.1038/s41467-025-57016-0

[4] Dawit H, Zhao Y, Wang J, Pei R. Advances in conductive hydrogels for neural recording and stimulation. Biomaterials Science. Royal Society of Chemistry. 2024;**12**:2786–2800

[5] Zhang P, Yang Y, Li Z, Xue Y, Wang F, Shan L, et al. Conducting hydrogel-based neural biointerfacing technologies. In: Advanced Functional Materials. Vol. **35**. Hoboken, New Jersey, U.S: John Wiley and Sons Inc; 2025

[6] Cheng S, Zhu R, Xu X. Hydrogels for Next Generation Neural Interfaces. Vol. 5, Communications Materials. Berlin, Germany: Springer Nature; 2024

[7] Lyu X, Hu Y, Shi S, Wang S, Li H, Wang Y, et al. Hydrogel Bioelectronics for Health Monitoring. Vol. **13**. Basel, Switzerland: Biosensors. Multidisciplinary Digital Publishing Institute (MDPI); 2023

[8] Ahmadi Seyedkhani S, Irajizad A, Mohammadpour R. Conductive Polymers and Hydrogels for Brain-Computer Interfaces. Berlin, Germany: Springer Nature; 2025. p. 117–175

[9] Hsieh JC, Li Y, Wang H, Perz M, Tang Q, Tang KWK, et al. Design of hydrogel-based wearable EEG electrodes for medical applications. Journal of Materials Chemistry B. 2022;**10**:7260–7280

[10] Martinelli E, Akouissi O, Liebi L, Furfaro I, Maulà D, Savoia N, et al. Applied sciences and engineering the e-flower: A hydrogel-actuated 3D MEA for brain spheroid electrophysiology. Science Advances. 2024;**10**. https://www.science.org

[11] Xue XY, Han L, Cai HQ, Zhang K, Sun ZC, Liu RP, et al. Conductive hydrogel-based neural interfaces: From fabrication methods, properties, to applications. Small Structures. 2025;**6**(8):2400696

[12] Wang L, Hajee Y, Frimat JP, Diba M, Savva A. Electrically active hydrogels based on PEDOT: PSS for neural cultures. Journal of Materials Chemistry C. 2025;(2):648–659. DOI: 10.1039/D5TC02708J

[13] Ahmed EM. Hydrogel: Preparation, characterization, and applications: A review. Journal of Advanced Research. 2015;**6**:105–121

[14] Hua J, Su M, Sun X, Li J, Sun Y, Qiu H, et al. Hydrogel-based bioelectronics and their applications in health monitoring. Biosensors. 2023;**13**:696

[15] Venkatadri TK, Ye C, Cohen T, Lin S. Torsion-mediated instabilities in confined elastic layers. Journal of the Mechanics and Physics of Solids. 2025;**198**:106064

[16] Sagdic K, Fernández-Lavado E, Mariello M, Akouissi O, Lacour SP. Hydrogels and conductive hydrogels for implantable bioelectronics. MRS Bulletin. 2023;**48**:495–505

[17] Wang S, Qian K, Mei Z, Zhu W, Zhou Z, Ye M, et al. High-strength conductive hydrogel fiber prepared via microfluidic technology for functionalized strain sensing. Macromolecular Rapid Communications. 2025. DOI: 10.1002/marc.202500222

[18] Hosseinzadeh S, Rezayat SM, Giaseddin A, Aliyan A, Soleimani M. Microfluidic system for synthesis of nanofibrous conductive hydrogel and muscle differentiation. Journal of Biomaterials Applications. 2018;**32** (7):853–861

[19] Bakarich SE, Gorkin R, Panhuis M, Spinks GM. Three-dimensional printing fiber reinforced hydrogel composites. ACS Applied Materials & Interfaces. 2014;**6**(18):15998–16006

[20] Yang J, Chen Y, Zhao L, Zhang J, Luo H. Constructions and properties of physically cross-linked hydrogels based on natural polymers. In: Patrick T. Mather, editor. Polymer Reviews. Vol. **63**. London, England: Taylor and Francis Ltd; 2023. p. 574–612

[21] Hu W, Wang Z, Xiao Y, Zhang S, Wang J. Advances in crosslinking strategies of biomedical hydrogels. Biomaterials Science. 2019;**7**:843–855

[22] Ho TC, Chang CC, Chan HP, Chung TW, Shu CW, Chuang KP, et al. Hydrogels: Properties and applications in biomedicine. Molecules. 2022;**27**:2902

[23] Zhong Y, Lin Q, Yu H, Shao L, Cui X, Pang Q, et al. Construction methods and biomedical applications of PVA-based hydrogels. Frontiers in Chemistry. 2024;**12**

[24] Priya AS, Premanand R, Ragupathi I, Bhaviripudi VR, Aepuru R, Kannan K, et al. Comprehensive review of hydrogel synthesis, characterization, and emerging applications. Journal of Composites Science. 2024;**8**:457

[25] Bashir S, Hina M, Iqbal J, Rajpar AH, Mujtaba MA, Alghamdi NA, et al. Fundamental concepts of hydrogels: Synthesis, properties, and their applications. Polymers. 2020;**12**:1–60

[26] Badali E, Hosseini M, Mohajer M, Hassanzadeh S, Saghati S, Hilborn J, et al. Enzymatic crosslinked hydrogels for biomedical application. Polymer Science - Series A. 2021;**63**: S1–22

[27] Agrawal A, Hussain CM. 3D-Printed hydrogel for diverse applications: A review. Gels. 2023;**9**:960

[28] Alahi MEE, Liu Y, Xu Z, Wang H, Wu T, Mukhopadhyay SC. Recent advancement of electrocorticography (ECoG) electrodes for chronic neural recording/stimulation. Materials Today Communications. 2021;**29**:102853

[29] Schalk G, Miller KJ, Anderson NR, Wilson JA, Smyth MD, Ojemann JG, et al. Two-dimensional movement control using electrocorticographic signals in humans. Journal of Neural Engineering 2008;**5**:75–84

[30] Hochberg LR, Serruya MD, Friehs GM, Mukand JA, Saleh M, Caplan AH, et al. Neuronal ensemble control of prosthetic devices by a human with tetraplegia. Nature 2006;**442**:164–71

[31] Chen S, Jiang D, Li M, Xuan X, Li H. Brain-computer interface and electrochemical sensor based on boron-nitrogen Co-Doped Graphene-diamond microelectrode for EEG and Dopamine detection. ACS Sensors. 2025;**10**(2):868–880

[32] Zhu Y, Chen B, Liu Y, Tan T, Gao B, Lu L, et al. Recent advances in conductive hydrogels for electronic skin and healthcare monitoring. Biosensors. 2025;**15**:463

[33] Ji Z, Li L, Zheng M, Ye X, Yan W, Wang Z, et al. Conductive Hydrogel-Enabled Electrode for Scalp Electroencephalography Monitoring. Small Methods. Hoboken, New Jersey, U.S: John Wiley and Sons Inc; 2025

[34] Vertes RP. Hippocampal theta rhythm: A tag for short-term memory. Hippocampus. 2005;**15**:923–935

[35] Xue H, Wang D, Jin M, Gao H, Wang X, Xia L, et al. Hydrogel electrodes with conductive and substrate-adhesive layers for noninvasive long-term EEG acquisition. Microsystems & Nanoengineering. 2023;**9**(1):79

[36] Butt H. Use of hydrogels in brain applications. Global Journal of Engineering Sciences. 2020;**4**(3)

[37] Merletti R. The electrode-skin interface and optimal detection of bioelectric signals. Physiological Measurement. 2010;**31**

[38] Zhang M, Qian B, Gao J, Zhao S, Cui Y, Luo Z, et al. Recent advances in portable dry electrode EEG: Architecture and applications in brain-computer interfaces. Sensors. 2025;**25**:5215

[39] Wang Z, Chen L, Chen Y, Liu P, Duan H, Cheng P. 3D printed ultrastretchable, hyper-antifreezing conductive hydrogel for sensitive motion and electrophysiological signal monitoring. Research. 2020;**2020**:426078

[40] Singh K, Lin CC, Huang WH, Lou LW, Chiueh H, Wang YH, et al. Ultrabioconformal, self-healable, and antioxidized polydopamine-inspired nanowire hydrogels enable resolving power in forehead and ear electroencephalograms for brain function assessment. ACS Applied Materials & Interfaces 2025;**17**:24887–24900

[41] Zhang Y, Hu Y, Xie B, Yang G, Yin Z, Wu H. Hoffmeister effect optimized hydrogel electrodes with enhanced electrical and mechanical properties for nerve conduction studies. Research. 2024;**7**:0453

[42] Wang D, Xue H, Xia L, Li Z, Zhao Y, Fan X, et al. A tough semi-dry hydrogel electrode with anti-bacterial properties for long-term repeatable non-invasive EEG acquisition. Microsystems & Nanoengineering. 2025;**11**(1):105

[43] Ren G, Zhang M, Zhuang L, Li L, Zhao S, Guo J, et al. MRI and CT compatible asymmetric bilayer hydrogel electrodes for EEG-based brain activity monitoring. Microsystems & Nanoengineering. 2024;**10**(1):156

[44] Liu Q, Yang L, Zhang Z, Yang H, Zhang Y, Wu J. The feature,

performance, and prospect of advanced electrodes for electroencephalogram. Biosensors. 2023;**13**:101

[45] Kaveh R, Schwendeman C, Pu L, Arias AC, Muller R. Wireless ear EEG to monitor drowsiness. Nature Communications. 2024;**15**(1):6520

[46] Li G, Wang S, Li M, Duan YY. Towards real-life EEG applications: Novel superporous hydrogel-based semi-dry EEG electrodes enabling automatically "charge-discharge" electrolyte. Journal of Neural Engineering. 2021;**18**(4):046016

[47] Han Q, Zhang C, Guo T, Tian Y, Song W, Lei J, et al. Hydrogel nanoarchitectonics of a flexible and self-adhesive electrode for long-term wireless electroencephalogram recording and high-accuracy sustained attention evaluation. Advanced Materials. 2023;**35**(12): 2209606

[48] Liang Q, Xia X, Sun X, Yu D, Huang X, Han G, et al. Highly stretchable hydrogels as wearable and implantable sensors for recording physiological and brain neural signals. Advanced Science. 2022;**9**(16):2201059

[49] Shen G, Gao K, Zhao N, Wang Z, Jiang C, Liu J. A fully flexible hydrogel electrode for daily EEG monitoring. IEEE Sensors Journal. 2022;**22**(13): S1–S22

[50] Ping H, Zhang Q, Chen X, Wang Y, Wu D, Wang J, et al. Epileptogenic zone-targeting nanoparticle drug delivery system for antiseizure medication. Materials Today Bio. 2025;**35**:102297

[51] Ahmadi Seyedkhani S, Irajizad A, Mohammadpour R, Taghipoor M, Vafaiee M. Novel brain-inspired hierarchical micro-nanostructured poly (3,4-ethylenedioxythiophene)/ polydopamine neural interface on titanium nitride electrodes for electrophysiological signal recording. ACS Applied Bio Materials. 2025;**8**:9332–9345

[52] Xiang Z, Yang L, Yu B, Zeng Q, Huang T, Shi S, et al. Recent advances in polymer-based thin-film electrodes for ECoG applications. Journal of Materials Chemistry B. 2024;**13**:454–471

[53] Tanwar A, Gandhi HA, Kushwaha D, Bhattacharya J. A review on microelectrode array fabrication techniques and their applications. Materials Today Chemistry. 2022;**26**:101153

[54] Shen W, Das S, Vitale F, Richardson A, Ananthakrishnan A, Struzyna LA, et al. Microfabricated intracortical extracellular matrix-microelectrodes for improving neural interfaces. Microsystems & Nanoengineering. 2018;**4**(1):30

[55] Kim Y, Alimperti S, Choi P, Noh M. An inkjet printed flexible electrocorticography (ECoG) microelectrode array on a thin parylene-C film. Sensors. 2022;**22**(3):1277

[56] Geramifard N, Lawson J, Cogan SF, Black BJ. A novel 3D helical microelectrode array for in vitro extracellular action potential recording. Micromachines (Basel). 2022;**13** (10):1692

[57] Tolstosheeva E, Gordillo-González V, Biefeld V, Kempen L, Mandon S, Kreiter AK, et al. A multi-channel, flex-rigid ECoG microelectrode array for visual cortical interfacing. Sensors (Switzerland) 2015;**15**:832–854

[58] Lee JY, Park SH, Kim Y, Cho YU, Park J, Hong JH, et al. Foldable three dimensional neural electrode arrays for simultaneous brain interfacing of cortical surface and intracortical multilayers. Npj Flexible Electronics. 2022;**6**(1):86

[59] Shin H, Jeong S, Lee JH, Sun W, Choi N, Cho IJ. 3D high-density microelectrode array with optical stimulation and drug delivery for investigating neural circuit dynamics. Nature Communications. 2021;**12**(1):492

[60] Choi Y, Park K, Choi H, Son D, Shin M. Self-healing, stretchable, biocompatible, and conductive alginate hydrogels through dynamic covalent bonds for implantable electronics. Polymers (Basel). 2021;**13**(7):1133

[61] Park S, Yuk H, Zhao R, Yim YS, Woldeghebriel EW, Kang J, et al. Adaptive and multifunctional hydrogel hybrid probes for long-term sensing and modulation of neural activity. Nature Communications. 2021;**12**(1):3435

[62] Lucherini L, Navello V, Akouissi O, Lacour SP, Amstad E. Direct laser writing of electronically conductive microstructures within soft hydrogel substrates. Materials Today Bio. 2025;**34**:102140

[63] Xie X, Xu Z, Yu X, Jiang H, Li H, Feng W. Liquid-in-liquid printing of 3D and mechanically tunable conductive hydrogels. Nature Communications. 2023;**14**(1):4289

[64] Wang F, Xue Y, Chen X, Zhang P, Shan L, Duan Q, et al. 3D printed implantable hydrogel bioelectronics for electrophysiological monitoring and electrical modulation. Advanced Functional Materials. 2024;**34**(21):2314471

[65] Athukorala SS, Tran TS, Balu R, Truong VK, Chapman J, Dutta NK, et al. 3D printable electrically conductive hydrogel scaffolds for biomedical applications: A review. Polymers. 2021;**13**:1–21

[66] Rahman S, Shon A, Joseph R, Pavlov A, Stefanov A, Namkoong M, et al. APPLIED SCIENCES AND ENGINEERING soft, stretchable conductive hydrogels for high-performance electronic implants. Science Advances. 2025;**11**(12):eads4415. https://www.science.org

[67] Oribe S, Yoshida S, Kusama S, Ichiro OS, Nakagawa A, Iwasaki M, et al. Hydrogel-based organic subdural electrode with high conformability to brain surface. Scientific Reports. 2019;**9**(1):13379

[68] Lee S, Park K, Kum J, An S, Yu KJ, Kim H, et al. Stretchable surface electrode arrays using an alginate/PEDOT:PSS-based conductive hydrogel for conformal brain interfacing. Polymers (Basel). 2023;**15**(1):84

[69] Chen Z, Zhang Y, Ding J, Li Z, Tian Y, Zeng M, et al. Hydrogel-based multifunctional deep brain probe for neural sensing, manipulation, and therapy. ACS Nano 2025;**19**:21600–21613

[70] Chenani H, Kammarchedu V, Asgharian H, Ebrahimi A. Neurochemical profiling in urine: Multiplexed detection of dopamine and serotonin using ML-integrated laser-induced graphene biosensors. Biosensors & Bioelectronics. 2025;**290**:117981

[71] Fan P, Wang Y, Dai Y, Jing L, Liang W, Lu B, et al. Flexible microelectrode array probe for simultaneous detection of neural discharge and dopamine in striatum of mice aversion system. Sensors and Actuators B: Chemical. 2023;**390**:133990

[72] Kimmel DW, Leblanc G, Meschievitz ME, Cliffel DE. Electrochemical sensors and biosensors. Analytical Chemistry. 2012;**84**:685–707

[73] VIEW - 2023 Huang - Emerging single-atom catalysts in electrochemical biosensing

[74] Rahman MM, Asiri AM. Introductory Chapter: Electrochemical Sensors Technology. In: Rahman MM, editor. Electrochemical Sensors Technology. Croatia: InTech; 2017

[75] Huang L, Zhou Y, Hu X, Yang Z. Emerging combination of hydrogel and electrochemical biosensors. Small. 2025;**21**(5):2409711

[76] Madadelahi M, Romero-Soto FO, Kumar R, Tlaxcala UB, Madou MJ. Electrochemical sensors: Types, applications, and the novel impacts of vibration and fluid flow for microfluidic integration. Biosensors & Bioelectronics. 2025;**272**:117099

[77] Liu J, Gao X, Ma W, Zhang L, Wen X, Zhai M, et al. Flexible hydrogel sensor based on MoS2 for highly selective dopamine detection against catecholamine cross-interference. Analyst. 2025;**151**(2):542–551

[78] Gu S, Zhang Z. Biocompatible AgInS2@hydrogel microelectrode with enhanced photoelectrochemical sensitivity for real-time in vivo dopamine monitoring. Analyst. 2025;**150**(11):2350–2355

[79] Wu G, Zhang ET, Qiang Y, Esmonde C, Chen X, Wei Z, et al. Long-term in vivo molecular monitoring using aptamer-graphene microtransistors. bioRxiv. 2023. DOI: 10.1101/2023.10.18.562080